THE PHYSICAL CHEMISTRY
OF DYE ADSORPTION

THE PHYSICAL CHEMISTRY OF DYE ADSORPTION

I. D. RATTEE   AND   M. M. BREUER

Department of Colour Chemistry, Leeds University, England

Unilever Research Laboratory, Isleworth, Middlesex, England

1974

ACADEMIC PRESS
LONDON AND NEW YORK

A Subsidiary of Harcourt Brace Jovanovich, Publishers

ACADEMIC PRESS INC. (LONDON) LTD.
24–28 Oval Road,
London, NW1 7DX

U.S. Edition published by

ACADEMIC PRESS INC.
111 Fifth Avenue,
New York, New York 10003

Library of Congress Catalog Card Number: 73–9474
ISBN: 0–12–582550–1

PRINTED IN GREAT BRITAIN BY
BUTLER & TANNER LTD., FROME AND LONDON

Introduction

The study of the physical chemistry of dye adsorption has concerned chemists connected with colour making and colour-using industries for many years. Prior to the sixteenth century, dyeing was a secret technology and very closely guarded. In that century a number of works were published giving details of procedures. The most important of these was produced by Rosetti (1548) who sought to release information "which has been imprisoned for a great number of years in the tyrannical hands of those who have kept it hidden". Some idea of the strength of commercial security of the time is given by the reaction to the first alum works being started at Guiseborough in Yorkshire in Elizabethan times. Sir Thomas Chaloner seduced workmen from the alum works at Tolfa in Italy because no one knew how to produce this, at that time, vital chemical for dyeing operations. The Pope was so enraged by this that he endeavoured to recall them by curses and anathemas!

One of the benefits employed by members of a profitable "mystery" is freedom from external criticism of standards. This freedom still enjoyed by the medical and legal professions was lost by dyers in the 17th century. The newly acquired scientific freedom of the time caused some investigations of dyeing by Boyle and by Hooke under the auspices of the Royal Society in the period 1664–1669 but these great men were not the first to find more tractable problems on which to concentrate. Thus Bancroft (1813) remarks that "from this time it does not appear that any thing considerable was done for nearly the space of a century by men of science in this kingdom towards improving the arts of dyeing".

In the same period in France, Colbert, in an early exercise in Consumer Protection turned his ministerial attention to dyeing. The result was in 1669 the "Instruction générale pour la teinture des Laines et pour la culture des drogues ou ingrédients qu'on emploie". This regulation required the establishment of a quality control system and sixty years later the post of Inspector

General of the Dyeing Industry was established with the assistance of the Académie Royale des Sciences. The appointment of scientists to this post led to the initiation of work designed to explain the dyeing process as well as to improve quality control. The strong links between research and practice then established have been maintained to the present day.

The first Inspector General, Dufay de Cisternay, carried out numerous experiments over many years which led to the conclusion that dye was taken up by physical rather than chemical processes and involved specific sites of attachment, (Dufay de Cisternay, 1737). Hellot who succeeded Dufay produced an alternative theory which postulated the movement of dye into pores which having been opened up by heat retained the dye as particles when the material was cooled, (Hellot, 1789). This theory explained the role of mordants to the satisfaction of the dyers of the time but did not explain why the dye should enter the pores in the first place. The beginnings of a concept involving a physical binding process on a molecular scale followed the work of a later Inspector General, Macquer. There were several claimants to the authorship of the concept and its origins are described with a disarming lack of modesty by Bancroft (1813). There is a strong likelihood that the significant author of this concept was Bergman in Sweden whose ideas were taken up and published by Berthollet (1791). There was much data published by Berthollet on affinities of dyes for substrate although it must be recognized that the values represent in modern terms little more than points on isotherms. In a sense this was recognized by Berthollet (1801) himself when the Law of Mass Action was formulated. It is of interest to note that it was virtually a direct application of Berthollet's Law over a century later by Langmuir that produced the isotherm equation so important in adsorption studies of all kinds.

The scientific study of dyeing was in its early days not only complex but also made extremely difficult by the restrictions of chemical knowledge and the absence of proper experimental techniques. Although chemical knowledge advanced and natural dyes were replaced by synthetic dyes of known structure many of the problems remained until this century. This is not to say that great strides were not made using simple methods of experiment. The first colour chemist in the modern sense was Böttiger who in 1888 developed the direct cotton colours and the theory underlying their structure as a consequence of a systematic study relating the chemical structure of dyes to their dyeing behaviour on cotton. However the application of physical chemistry to dye adsorption problems was at a fairly crude level until the 1930's when S. M. Neale began his studies with dye interactions with cellulose. Later in the 1940's Rideal turned his attention to the thermodynamics of dye adsorption and his contribution was seized upon by a group of investigators under T. Vickerstaff who began to establish the subject as one seriously attracting physical chemists. Vickerstaff's book *The Physical Chemistry of Dyeing* published in the 1950's was the first book in English on the topic and had a very great influence both in raising the intellectual standards of technologists and attracting the interest of physical chemists. Vickerstaff's book still remains the only available reference work and but while it is a valuable source of information it does not reflect many of the changes of emphasis and understanding which have taken place over the past two decades. At the present time interest in adsorption at membrane surfaces extends far beyond textile colouration and the physical chemists concerned with the physical interactions of fairly complex molecules with synthetic and biopolymers are a rapidly growing band.

One of the difficulties of subjects born in technology is that their literature and philosophy reflects the divisions of technological operations. Thus traditionally and even in Vickerstaff's time the behaviour of dyes for cotton and dyes for wool is seen as being basically as well as superficially different. In the present work an attempt has been made to show the unity of the dye adsorption process whatever the chemical nature of the substrate whether it be a textile or biopolymer. It is not always realised how complex the "simple" approach of the technologist actually is and how the failure to analyse deeply leads to over complication in description. Nearly two centuries ago Henry (1740) another early student of dyeing theory remarked that "though long experience may establish a number of facts, yet if the rationale of the manner by which they are produced be not understood, misapplications are liable to be made; similar practices are pursued where the cases differ essentially; and improvements are attempted at hazard, and often on false principles".

The subject of the physical chemistry of dye adsorption is very wide indeed and covers a vast literature. No attempt has been made in the present work to present an exhaustive account of everything that has been done. On the other hand the present state of thinking in relation to the topic is presented so that the researcher entering the field or the student learning about it may acquire the necessary critical equipment.

The subject inevitably touches upon two topics of wide interest namely intermolecular forces and adsorption. No attempt is made to cover these very considerable subjects exhaustively and discussion is restricted as much as possible to the context of dye adsorption. Thus direct references to the literature are few and the reader is recommended to read further for a more general background. On the other hand other topics even when of general significance such as diffusion are considered in detail in relation to the established literature. This differentiation is, to the authors at any rate, reasonable and avoids excessive repetition of what may be read elsewhere.

REFERENCES

Bancroft, E. (1813) *The Philosophy of Permanent Colours*, London, Vol. 1, pp. XIX–XI.

Berthollet, C. L. (1791) *Éléments de l'art de la teinture*, Paris, Part 1.

Berthollet, C. L. (1803) *Essai de statique chimique*, Paris.

Dufay de Cisternay, G. F. (1737) *Mémoires de l'Académie des Sciences*, Paris, p. 253.

Hellot, J. (1789) *The Art of Dyeing Wool, Silk and Cotton*. Translated anonymously. R. Baldwin, London, p. 18.

Henry, T. (1790) *The Nature of Colouring Matters*.

Rosetti, G. (1548) *The Phictho*. Trans. S. M. Edelstein and H. G. Borghetty, M.I.T. Press, Cambridge, Mass.

Contents

Introduction

Chapter 4   Dyes in solution

Chapter 5   The adsorption of dyes by
proteins and polyamides

Chapter 6   The adsorption of dyes by
cellulosic substrates

Chapter 7   The adsorption of non-ionic
disperse dyes

Chapter 8   Fibre-reactive dyes

Chapter 9   Problems of dyeing kinetics—
specific situations

Chapter 10   Relating theory and practice

Glossary of Symbols

a   thermodynamic activity; fibre radius; expansion coefficient

A^1 frequency factor

b   specific surface energy

c, C   concentration; molarity; capacitance; partition coefficient

∇c concentration gradient

D   diffusion coefficient; dye concentration

$\bar{D}$ integral diffusion coefficient

D^* self diffusion coefficient

d   dipole moment; monomer to dimer ratio in stacked adsorption

E   surface energy; potential gradient; activation energy; dyebath exhaustion

e   elementary charge $(1{\cdot}6021 \times 10^{-19}\,C)$

F   Faraday constant $(9{\cdot}6487 \times 10^4\,C\,mol^{-1})$; surface area

f   free volume fraction; frictional coefficients; vector of external forces; activity coefficient

G   Gibbs' free energy (H–TS)

G^* Gibbs' free energy for a transitional stage

H   enthalpy

h   Planck's constant

I   electric current; ionic strength; ionisation potential

J   Bessel function

J_i flux

K   intrinsic partition coefficient; equilibrium constant; stacking probability, ratio

k Boltzmann's constant $(1{\cdot}3805 \times 10^{-23}\,JK^{-1})$

k   adsorption constant; reaction rate constant

L   mean free path

L_{ij} phenomenological coefficient

M   molecular weight; uptake

m   mass; mobility

N   Avogadro's number

n   number of molecules; stirring rate (r.p.m.); refractive index

P   probability

P_1 pressure; ratio of partial molar volumes

Q   heat (exchanged with surroundings); surface area; amount

Δq — material flow

R — resistance; gas constant ($8 \cdot 3143 \, JK^{-1} \, mol^{-1}$)

R^1 — reactivity ratio

R_e — Reynolds number

r — radius; distance; liquor ratio

S — entropy; surface area; concentration of surface adsorption sites

T — temperature ($°C$); time of half levelling

T_g — glass transition temperature

t — time

t_d — transport number

U — internal energy; velocity; potential energy

V — volume; velocity; mobility

V_f — free volume

$\bar{V}$ — partial molal volume

x, X — conjugate force; relative uptake; mole fraction

Z — valency or ionic charge

α — thermal expansion coefficient; diameter; polarisability

Γ — adsorbed surface concentration

γ — interfacial or surface tension

ϑ — Nerns boundary layer thickness

σ — density; equilibrium partition coefficient; tortuosity factor; electrical charge density

ζ — dielectric constant; molar extinction coefficient

ε' — elastic modulus

ε'' — storage or viscous modulus

η — viscosity

θ — radius vector angle; degree of surface site occupation

κ — reciprocal ion atmosphere radius or double layer thickness

Λ — limiting conductivity

λ — partition coefficient; extension ratio; length of displacement

μ — chemical potential

$\tilde{\mu}$ — electro-chemical potential

v — kinematic viscosity; frequency; frequency factor

ξ — frictional coefficient

τ — relaxation or dwell time; time of half dyeing

π — molecular interaction energy; potential energy

ϕ — dissipation function (rate of entropy change)

χ — interaction parameter

ψ — electrical potential

ω — circular frequency

Chapter 1

Binding Forces

I   INTRODUCTION

The adsorption process involves the selective uptake of a substance from an external phase into a phase provided by the adsorbant. Thus, when an adsorbant is immersed in a solution adsorption takes place when the solute molecules transfer from the solution to the adsorbant leading to a change in the concentration of the solution. The process differs from imbibition which involves the simple uptake of unchanged solution. For a transfer such as described to occur spontaneously it must result in the system reaching a *state of lower energy*, or higher stability, as a consequence. The energy released may be regarded as *free energy* and its creation provides a driving force leading to adsorption. Clearly the reverse process of desorption requires that energy be put back into the system so that free energy creation measures the resistance of the system to change, i.e. desorption, and manifests itself as a *binding force* between the adsorbate and adsorbant. The factors in the system which lead to free energy creation may be of several kinds and can be considered in terms of different kinds of binding force. In the context of dye adsorption the binding forces are dye-substrate bonds and they are *reversible*, i.e. due to physical interactions between dye and substrate, or *irreversible*, i.e. due to formation of covalent links as a result of chemical interaction. The consideration of the latter kind of binding force may be regarded as a special case and attention is directed to Chapter 8. Reversible binding forces which are important either in fact or theory in relation to dye substrate interactions arise from electrical interactions involving ionized change centres, or dipoles of a permanent or induced nature. They are most usefully considered under the following headings:

(a)   coulombic interactions
(b)   dipole interactions or perturbation forces
(c)   hydrophobic interactions.

II   COULOMBIC INTERACTIONS.

Where both the adsorbant and the adsorbate bear a charge, i.e. both are ionized, then attractive forces inversely related to the distance between the charge centres can operate. Dyes may be anionic and cationic and adsorbing substrates such as proteins, carbohydrates and polyamides may be charged also. Even if the adsorbant bears no formal charge before adsorption of a dye ion, the process of adsorption produces a surface charge and the surface potential is determined by the adsorbed dye ions. Coulombic interactions can either favour or disfavour dye adsorption so that they may be regarded as binding or antibinding forces as appropriate.

The presence of a charge on an adsorbing surface confers upon it an electrical potential which simultaneously attracts oppositely charged (counter-) ions and repels similarly charged (co-) ions. This is represented schematically in Fig. 1.1.

As the distance from the charged surface increases the effect of the surface potential becomes less. Thus the surface potential is only apparent over a limited distance in a practical sense and within this limited distance all of the imbalance of ionic concentrations near the surface is combined. This means that near to the surface exists a diffuse electrical layer and since within it there will be a charge separation it is termed an *electrical double layer*. It

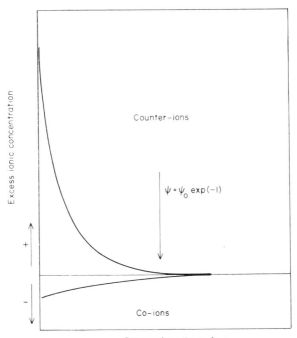

Fig. 1.1.   The distribution of ions near a charged surface as a function of distance.

may be treated theoretically in various ways based on two dominant approaches due to Gouy and Donnan.

The Gouy treatment of the problem provides the basis for the consideration of the problem of the distribution of ions relative to other ions in solution in accordance with the Debye–Hückel–Onsager theory. The Donnan treatment which was developed at about the same time provides an alternative and often very useful simple approach to the problem.

A   The Gouy treatment of the electrical double layer

The Gouy treatment is based on certain assumptions about the surface and the ions involved:

(a)   that the charged surface is planar
(b)   that the charges on the surface are not localized
(c)   that counter-ions and co-ions do not interact
(d)   that counter-ions have zero size and do not penetrate the surface
(e)   that the effects of the surface potential are only significant perpendicular to the surface.

The situation analysed by the Gouy treatment is expressed diagramatically in Fig. 1.2.

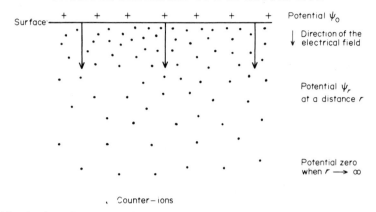

Fig. 1.2. A schematic representation of the distribution of ions near a charged surface in accord with the Gouy model.

The potential energy, U, of an ion in a potential field is given by

$$U = ze\psi \qquad (1.1)$$

where z is the charge on the ion, e is the electronic charge and ψ is the potential at the point where the ion exists.

The probability of finding an ion at some particular point relative to a potential surface involves the potential energy and the Boltzmann distribution factor so that the probability p is given by

$$p = \exp(ze\psi/kT) \qquad (1.2)$$

in which k is the Boltzmann distribution constant and T is the temperature.

The electrical force factor may increase or reduce the probability because it may be attractive or repulsive. The probability term may be replaced by appropriate concentration terms to give two Nernst distribution equations for the two kinds of ion. Thus assuming a positive surface and the same numerical value of z for the two ions,

$$C^+ = C_0 \exp(-ze\psi/kT) \qquad (1.3)$$

$$C^- = C_0 \exp(+ze\psi/kT) \qquad (1.4)$$

where C^+, C^- and C_0 are the cationic, anionic and general ionic concentrations respectively.

The gradient of the electrical potential at a distance r from the surface is given by Poisson's equation,

$$\frac{\partial^2\psi}{\partial r^2} = -\frac{4\pi\rho}{\varepsilon} \qquad (1.5)$$

in which ε is the dielectric constant of the medium and ρ is the net charge density, i.e. the excess charge in the present case. Thus

$$\rho = -ze\left(C^- - C^+\right) \tag{1.6}$$

Combining equations 1.3, 1.4 and 1.6 gives

$$\rho = -zeC_0\left[\exp\left(ze\psi/kT\right) - \exp\left(-ze\psi/kT\right)\right]$$
$$= -2zeC_0\sinh\left(ze\psi/kT\right) \tag{1.7}$$

Substituting equation 1.7 in 1.5 gives

$$\frac{\partial^2\psi}{\partial r^2} = \frac{8\pi zeC_0}{\varepsilon}\sinh\left(\frac{ze\psi}{kT}\right) \tag{1.8}$$

Commencing the integration of equation (1.8)

$$\frac{1}{2}\left|\frac{\partial\psi}{\partial r}\right|^2 = \left(\frac{8\pi C_0 kT}{\varepsilon}\right)\cosh\left(\frac{ze\psi}{kT}\right) + A \tag{1.9}$$

where A is a constant of integration.

Since $\partial\psi/\partial r = \psi = 0$ when $r = \infty$ by definition,

$$A = \frac{8\pi C_0 kT}{\varepsilon} \tag{1.10}$$

and

$$\frac{1}{2}\left|\frac{\partial\psi}{\partial r}\right|^2 = \left(\frac{8\pi C_0 kT}{\varepsilon}\right)\left[\cosh\left(\frac{ze\psi}{kT}\right) - 1\right] \tag{1.11}$$

or

$$\frac{1}{2}\left|\frac{\partial\psi}{\partial r}\right|^2 = \left(\frac{16\pi C_0 kT}{\varepsilon}\right)\sinh^2\left(\frac{ze\psi}{2kT}\right) \tag{1.12}$$

Completing the integration and rearranging gives

$$\psi_r = \left(\frac{2kT}{ze}\right)\ln\frac{\left[1 + A\exp\left(-\kappa r\right)\right]}{\left[1 - A\exp\left(-\kappa r\right)\right]} \tag{1.13}$$

in which A is a constant of integration and $\kappa = \left[\dfrac{8\pi C_0 z^2 e^2}{\varepsilon kT}\right]^{1/2}$

Doubly differentiating equation (1.13) and substituting $r = 0$ gives

$$\frac{\partial^2\psi}{\partial r^2} = -\left(\frac{2kT}{ze}\right)\left(\frac{2Ak}{(1 - A^2)}\right) \tag{1.14}$$

$$= \left|\frac{-4\pi\rho}{D}\right|_{r=0} \quad \text{(equation 1.5)}$$

Thus

$$A = \frac{\left(1 \pm \sqrt{1 + 4\alpha^2}\right)}{2\alpha} \text{ where } \alpha = \pi\rho_0\left(8\pi C_0 zkT\varepsilon\right)^{-1/2} \tag{1.15}$$

where ρ_0 is ρ at $r = 0$.

The exact solution for ψ_0 can be obtained by putting $r = 0$ in equation (1.13) and using equation (1.15) to give

$$\sinh \frac{(ze\psi_0)}{(2kT)} = \left(\frac{\pi\rho_0^2}{2C_0 kT\varepsilon}\right)^{1/2}$$ (1.16)

so that rearranging gives

$$\psi_0 = \frac{(2kT)}{(ze)}\sinh^{-1}\frac{\sigma}{I^{1/2}}\sqrt{\frac{\pi}{2kT\varepsilon}}$$ (1.17)

since C_0 being the general ionic concentration equals the ionic strength I and the charge density ρ_0 at $r = 0$ equals the surface charge density σ.

Another way of dealing with the situation is to express ψ_r and ψ_0 in terms of

$$y = \frac{ze\psi_r}{kT} \text{ and } y_0 = \frac{ze\psi_0}{kT}$$ (1.18)

Substituting these values in equation (1.13) gives

$$\frac{ykT}{ze} = \left(\frac{2kT}{ze}\right)\ln\left[\frac{(1+Ae^{-\kappa r})}{(1-Ae^{-\kappa r})}\right]$$ (1.19)

and

$$\frac{y_0 kT}{ze} = \left(\frac{2kT}{ze}\right)\ln\left[\frac{(1+A)}{(1-A)}\right]$$ (1.20)

From equation (1.20)

$$A = \frac{[\exp(y_0/2) \ 1]}{[\exp(y_0/2)+1]}$$ (1.21)

so that substituting equation (1.21) in equation (1.19) gives

$$\exp(y/2) = \frac{\exp(y_0/2)+1+(\exp(y_0/2)-1)\exp(-\kappa r)}{\exp(y_0/2)+1-(\exp y_0/2-1)\exp(-\kappa r)}$$ (1.22)

When $y_0 \ll 1$

$$y = y_0 \exp(-\kappa r)$$

$$\psi = \psi_0 \exp(-\kappa r)$$ (1.23)

i.e. the potential has fallen to $1/e$ of the value at the surface when $\kappa = 1/r$. The distance $r = 1/\kappa$ corresponds to the ion atmosphere radius of the Debye–Hückel–Onsager theory and is generally taken as the thickness of the electrical double layer. The degree to which this is true is indicated in Fig. 1.1 by the line showing $\psi = \psi_0/\exp(-1)$. The area of the charged surface times $1/\kappa$

will correspond to the volume of the electrical double layer and from the definition of κ as

$$\left[\sqrt{\frac{8\pi C_0 z^2 e^2}{kT\varepsilon}}\right]$$

the volume V is expected to vary as follows,

$$V = \alpha I^{-1/2} \qquad \text{where } \alpha = \left(\frac{\text{Area}}{ze}\right)\left(\frac{\varepsilon kT}{8\pi}\right)^{1/2} \tag{1.24}$$

$$= \beta z \qquad \text{where } \beta = \left(\frac{\text{Area}}{e}\right)\left(\frac{\varepsilon kT}{8\pi I}\right)^{1/2} \tag{1.25}$$

$$= \gamma (T)^{1/2} \qquad \text{where } \gamma = \left(\frac{\text{Area}}{ze}\right)\left(\frac{\varepsilon kT}{8\pi I}\right)^{1/2} \tag{1.26}$$

The assumptions of the Gouy theory of the double layer are open to question since ions clearly possess a finite size and will interact with other ions. Another problem is that when calculating surface potentials a value has to be ascribed to ε. Normally the bulk dielectric constant is employed but this may well be seriously in error due to dielectric saturation in the powerful electrical field which can exceed 10^5 volts . cm^{-1}. In addition the solvent may be polarized near the surface to give abnormal effects. In practice the errors are largely self cancelling over a wide range of real conditions including those employed in dye absorption studies. However the errors can be very considerable in some circumstances.

B   The Donnan membrane theory

The Donnan membrane theory of the electrical double layer is mathematically more simple although its assumptions are rather more complex than the Gouy treatment. If the decay of potential with distance from the surface is represented schematically by the line AB in Fig. 1.3, then the integration of $\partial\psi/\partial r$ which is involved in the Gouy treatment gives the area ABC. If it is now postulated that a conceptual potential ψ_{Donnan} exists which remains constant with r and then falls to zero at some characteristic distance, it is clear that a value of ψ_{Donnan} and the characteristic distance can be chosen so as to give the same integrated area as ABC. One such value is shown as PQRC. The Donnan membrane theory treats the electrical double layer in this way, i.e. it presupposes a constant potential, the Donnan potential, which remains constant across the thickness of a conceptual double layer and then falls to zero. Thus QR in Fig. 1.3 represents an outer limit of the double layer across which there is a marked change of potential. This is regarded by Donnan as a membrane permeable

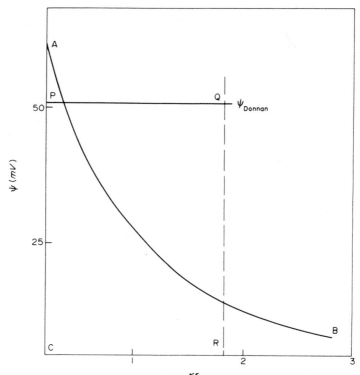

Fig. 1.3. Schematic representation of the Donnan membrane model for the surface potential.

to solute ions through which the fixed charges in the surface cannot escape. From the point of view of the calculation of distribution of ions in the system, the Donnan model should be simpler mathematically than the Gouy model. From the point of view of calculating the surface potential it will tend to underestimate the surface potential i.e. $\psi_{Donnan} < \psi_0$. The distribution of ions and surface charges in the Donnan model is normally presented schematically as follows

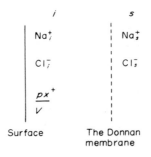

The Donnan membrane corresponds to the line QR in Fig. 1.3. The phase

on the surface side of the membrane is indicated by a suffix i and that on the solution side by the suffix s. The term px^+/V represents the surface charges expressed as a concentration in the surface phase i. If we have p charges of type x^+ per unit mass of adsorbant, and the volume of the Donnan double layer per unit mass is V, then the fixed charges expressed as a concentration in the surface phase is px^+/V.

At equilibrium the ions of solute will be distributed so that the electrochemical potential of either species will be the same in each phase, i.e.

$$\bar{\mu}_i = \bar{\mu}_s \text{ for all species} \tag{1.27}$$

for the positive surface indicated in the diagram,

$$(\bar{\mu}_i)_{\text{Na}+} = (\bar{\mu}_0)_{\text{Na}_i+} + kT \ln [\text{Na}_i^+]\gamma_{\text{Na}_i+} + e\psi_{\text{Donnan}} \tag{1.28}$$

$$(\bar{\mu}_s)_{\text{Na}+} = (\bar{\mu}_0)_{\text{Na}_s+} + kT \ln [\text{Na}_s^+]\gamma_{\text{Na}_s+} \tag{1.29}$$

$$(\bar{\mu}_i)_{\text{Cl}-} = (\bar{\mu}_0)_{\text{Cl}_i-} + kT \ln [\text{Cl}_i^-]\gamma_{\text{Cl}_i-} - e\psi_{\text{Donnan}} \tag{1.30}$$

$$(\bar{\mu}_s)_{\text{Cl}-} = (\bar{\mu}_0)_{\text{Cl}_s-} + kT \ln [\text{Cl}_s^-]\gamma_{\text{Cl}_s-} \tag{1.31}$$

If we now assume that the phases i and s are of the same kind i.e. ideal aqueous solutions, the standard electrochemical potential for either species will be the same in the two phases i.e. $(\bar{\mu}_0)_{\text{Na}_i+} = (\bar{\mu}_0)_{\text{Na}_s+}$ etc. Thus when equations 1.28 and 1.29 or 1.30 and 1.31 are combined, in view of equation (1.27)

$$\ln \left\{ \frac{[\text{Na}_i^+]}{[\text{Na}_s^+]} \right\} = -\left(\frac{e\psi_{\text{Donnan}}}{kT} \right) - \ln \frac{(\gamma_{\text{Na}_i+})}{(\gamma_{\text{Na}_s+})} \tag{1.32}$$

$$\ln \left\{ \frac{[\text{Cl}^{-1}]}{[\text{Cl}_i^-]} \right\} = -\left(\frac{e\psi_{\text{Donnan}}}{kT} \right) - \ln \frac{(\gamma_{\text{Cl}_s-})}{(\gamma_{\text{Cl}_i-})} \tag{1.33}$$

Assuming the same degree of non-ideality in both phases; that is, neglecting the activity terms,

$$\psi_{\text{Donnan}} = \left(\frac{kT}{ze} \right) \ln \left\{ \frac{[\text{Na}_s^+]}{[\text{Na}_i^+]} \right\} \cdot \left\{ \frac{[\text{Cl}_i^-]}{[\text{Cl}_s^-]} \right\} \tag{1.34}$$

$$\text{Also } \frac{[\text{Na}_i^+]}{[\text{Na}_s^+]} = \frac{[\text{Cl}_s^-]}{[\text{Cl}_i^-]} = \exp\left(-\frac{e\psi_{\text{Donnan}}}{kT} \right) = \text{a constant} \tag{1.35}$$

Equation 1.35 is particularly valuable as will be seen. The constant is sometimes termed the Donnan distribution coefficient.

There is another important feature of the system in that at equilibrium electrical neutrality should exist so that

$$[\text{Na}_i^+] + p\left(\frac{[x^+]}{V} \right) = [\text{Cl}_i^-] \tag{1.36}$$

Combining this equation with (1.35) gives

$$[Na_i^+] - \left\{ \frac{[Na_s^+][Cl_s^-]}{[Na_i^+]} \right\} + \left\{ \frac{[p\,x^+]}{V} \right\} = 0 \qquad (1.37)$$

$$[Cl_i^-] - \left\{ \frac{[Na_s^+][Cl_s^-]}{[Cl_i^-]} \right\} - \left\{ p\,\frac{[x^+]}{V} \right\} = 0 \qquad (1.38)$$

The two quadratic equations in $[Na_i^+]$ and $[Cl_i^-]$ can be readily solved to give

$$[Na_i^+] = \left\{ \frac{p^2[x^+]^2}{4V^2} + [Na_s^+][Cl_s^-] \right\}^{1/2} - \frac{p[x^+]}{2V} \qquad (1.39)$$

$$[Cl_i^-] = \left\{ \frac{p^2[x^+]^2}{4V^2} + [Na_s^+][Cl_s^-] \right\}^{1/2} + \frac{p[x^+]}{2V} \qquad (1.40)$$

Substituting the two solutions into equation (1.34) gives

$$\psi_{\text{Donnan}} = \frac{kT}{e} \ln \left\{ \frac{p[x^+]}{2VI} + \left[\frac{p^2[x^+]^2}{4I^2V^2} + 1 \right]^{1/2} \right\} \qquad (1.41)$$

in which $I = [Na_s^+] = [Cl_s^-]$

The thickness of the Donnan double layer will be subject to the same effects as in the case of the diffuse double layer of Gouy. Thus the volume of the Donnan double layer will vary in the same way as κ^{-1}. Thus the value of ψ_{Donnan} and the Donnan distribution coefficient will vary with ionic strength, ionic charge and temperature.

C Potential determination

So far the distribution of simple ions has been considered. With such ions only force attracting them into the surface phase is electrostatic. However if other forces are in operation so that specific adsorption of a kind yet to be considered occurs then the adsorbed ions will become, as it were, part of the surface and will contribute to the surface potential. In addition to this possibility, the charge centres in the surface may be weakly basic or acidic with the consequence that the net charge density will depend upon the acid-base equilibria operating in the surface phase. The surface potential will consequently depend upon the pH in the surface phase. These two factors may be considered separately although in practice they are likely to operate together.

(1) Specific adsorption effects

Several cases are known in which in the presence of increasing concentrations of certain ions the charge on an adsorbing surface will reduce and then reverse. However according to the Gouy equation, the potential does not change sign as I increases but only tends towards zero. Gilbert and Rideal have argued that it is necessary where the ions may be adsorbed to introduce an additional energy term to allow for the specific interaction. Taking a specific example from their argument i.e. the adsorption of HCl by wool this gives

$$\ln\left[\frac{\theta}{(1-\theta)}\right] = \ln I + \left[\frac{(\lambda p + W)}{kT}\right] - \left(\frac{e\psi}{kT}\right) \tag{1.42}$$

in which θ is the fraction of carboxyl groups back titrated by hydrogen ions, and $(\lambda p + w)$ is the total interaction energy of specific interactions with hydrogen ions. There are a great many objections to this treatment which will be considered at a later stage but the equation does in a broad sense agree with the experiment.

An earlier consideration of the same problem by Stern allows not only for specific adsorption energy, but also non-zero size of the counter-ions. Stern's equation is

$$n_c = c\left\{\frac{n_s - n_c}{(1000/M)}\right\} \exp\left(-ze\psi_\delta + \lambda_p + w\right)/kT \tag{1.43}$$

Here n_c is the number of counter-ions in the surface phase per square cm of surface, their charge is z and n_s is the maximum capacity of the surface for counter-ions, M is the molecular weight of the solvent, ψ_δ is the potential at the plane of the first layer of counter-ions on the surface, i.e. the counter-ions are assumed to form a layer of thickness δ. The term $(-ze\psi_\delta + \lambda_p + w)/kT$ is the sum of interacting energies, coulombic and otherwise.

The term $(n_s - n_c)$ allows for the finite size of the counter-ions. For a single ionic species, n_c/n_s may be expressed as θ, the fractional occupation of the surface. Taking z as unity, equation (1.43) can now be rewritten as

$$\ln\frac{\theta}{1-\theta} = \ln I + \left[\frac{\lambda_p + w}{kT}\right] - \left(\frac{e\psi_\delta}{kT}\right) + \ln\frac{M}{1000} \tag{1.44}$$

which is almost identical with the equation of Gilbert and Rideal (1.42).

(2) Acid-base equilibria and internal pH effects

Clearly the coulombic attraction or repulsion of protons by a charged surface will lead to a surface phase which has a different pH from that of the bulk solution. The simple Nernst equation for a negative surface gives

$$[H_i^+] = [H_s^+] \exp\left(e\psi/kT\right) \tag{1.45}$$

Consequently $pH_i = pH_s - ze\psi/2.303\,kT$. $\tag{1.46}$

This situation can have many repercussions. If the negative surface in the above example arises from weakly acid groups e.g. carboxyl groups in the surface, the degree of ionization will depend upon $[H_i^+]$ and the proton will be potential determining. In addition proton catalysed reactions such as polymerization, hydrolysis etc. will proceed at rates determined by $[H_i^+]$. Thus calculation of the so-called internal pH is frequently important and in this respect the Donnan membrane equations are of very great value. This is best shown by an example.

For a surface containing carboxyl groups in the presence of saline HCl, the ionic balance equation gives

$$[H_i^+] + [Na_i^+] = [Cl_i^-] + [COO_i^-] \tag{1.47}$$

The Donnan distribution equation gives

$$\left\{\frac{[H_s^+]}{[H_i^+]}\right\} = \left\{\frac{[Na_s^+]}{[Na_i^+]}\right\} = \left\{\frac{[Cl_i^-]}{[Cl_s^-]}\right\} \tag{1.48}$$

The acid/base equilibrium equation gives

$$\frac{[COO_i^-][H_i^+]}{[COOH_i]} = K_c \quad \text{and} \quad [COO_i^-] = \frac{K_c[\bar{C}]}{K_c + [H_i^+]} \tag{1.49}$$

in which $[\bar{C}]$ equals $[COO_i^-] + [COOH_i]$.

Combining (1.47), (1.48) and (1.49) gives

$$[H_i^+] + \frac{[H_i^+][Na_s^+]}{[H_s^+]} = \frac{[H_s^+][Cl_s^-]}{[H_i^+]} + \frac{K_c[\bar{C}]}{K_c + [H_i^+]} \tag{1.50}$$

This may be expanded to give a cubic equation in $[H_i^+]$,

$$[H_i^+]^3 + [H_i^+]^2 K_c - [H_i^+][H_s^+]\left\{[H_s^+] + K_c\frac{[\bar{C}]}{I}\right\} - K_c[H_s^+]^2 = 0 \tag{1.51}$$

since $[Na_s^+] + [H_s^+] = [Cl_s^-] = I$

If in addition to hydrochloric acid and sodium chloride, the sodium salt of a dye was present then equation (45) becomes,

$$[H_i^+] + [Na_i^+] = z[D^{z-}] + [Cl_i^-] + [COO_i^-] \tag{1.52}$$

which gives, assuming all the dye taken up is adsorbed on the adsorbing surface,

$$z[D^{z-}] = \frac{[H_s^+]I}{[H_i^+]} + \frac{K_c[\bar{C}]}{K_c + [H_i^+]} - \frac{[H_i^+]I}{[H_s^+]} \tag{1.53}$$

If the acid groups on the dye molecules are affected by the surface acidity then there are present a number of potential determining factors. Allowance can be made for these without difficulty. If $z = 3$ and the three dissociation constants are K_1, K_2 and K_3, equation (1.53) becomes,

$$\frac{[H_i^+]I}{[H_s^+]} = \frac{K_c}{K_c + [H_i^+]} + \frac{[H_s^+]I}{[H_i^+]}$$

$$+ 3[D^{3-}]\left[1 + \frac{[H_i^+]}{K_1} + \frac{[H_1^+]^2}{K_1 K_2} + \frac{[H_1^+]3}{K_1 K_2 K_3}\right]^{-1}$$

$$+ 2[D^{3-}]\left[1 + \frac{K_1}{[H_i^+]} + \frac{[H_i^+]}{K_2} + \frac{[H_1^+]^2}{K_2 K_3}\right]^{-1}$$

$$+ [D^{3-}]\left[1 + \frac{K_1 K_2}{[H_i^+]^2} + \frac{K_2}{[H_i^+]} + \frac{[H_i^+]}{K_3}\right]^{-1} \tag{1.54}$$

This quintic equation in $[H_i^+]$ is again solved fairly readily.

The general method of calculation in any case is to derive equations of the type of 1.47 and 1.48 followed by subsidiary equations of the type of 1.49 and then solve all equations simultaneously.

One important problem is the assignment of a concentration value to the fixed charges. These are generally known from analysis in terms of g. equivalent/kg and this can be transformed to a concentration only by a knowledge of the double layer volume per unit weight, V. This in turn requires the available surface area and the double layer thickness to be fixed. The thickness term can be ascribed the value $1/\kappa$ i.e. $\sqrt{(\varepsilon k T/8\,Iz^2 e^2)}$ and the available surface area measured by appropriate means. The term $1/\kappa$ is the ion atmosphere radius of the Debye–Hückel theory. Strictly speaking the definition of $1/\kappa$ given above applies only to symmetrical electrolytes, i.e. $1:1$, $2:2$ etc. Where $|z^+| \neq |z^-|$ then the correct term for κ is

$$\kappa = \left\{(4\pi e^2/\varepsilon k T)\sum_i n_i z_i^2\right\}^{1/2} \tag{1.55}$$

which gives the correct value for $I\,(=\frac{1}{2}\sum_i n_i z_i^2)$.

This approach has been taken in some recent studies of cellulose dyeing to be considered later.

D   A comparison of the Gouy and Donnan models with practice

A surface film of an appropriate ester (e.g. cholesterol formate) will hydrolyse at a rate determined by the surface concentration of protons. Such a film can be given a charge by the addition of a known amount of a fatty acid or base so that the surface potential may be calculated by the Gouy or Donnan methods. Calculation of the local hydrogen ion concentration will enable practical values to be compared with predicted values. This kind of experiment shows that broadly speaking the Gouy equations are more to be relied upon at low ionic strength and fairly low charge densities but the Donnan equations are better at high values of I and σ.

Sumner has followed internal pH charges using a covalently bound indicator in a cellophane film and has compared the results with those predicted using the Donnan equations. Excellent agreement was obtained.

E   Coulombic interactions and the adsorption of ions

It should be clear that three main coulombic effects will be encountered in dye adsorption.

(*a*)   The presence of acid may determine the potentials of the adsorbing phase to make it more or less attractive to dye ions e.g. proteins contain positive (NH_3^+) and negative (COO^-) groups. In acid the carboxyl groups are back titrated giving the fibre a positive charge favourable to the uptake of dye anions. Cellulose contains hydroxyl groups which in the presence of alkali may ionize giving the fibre a negative charge repelling dye anions and preventing their absorption.

(*b*)   Dye ions will be adsorbed by uncharged substrates to a very limited degree since the first few molecules taken up will confer on the fibre a charge repelling other similarly charged ions. This effect will be most marked in relatively hydrophobic fibres since the volume of water taken up broadly corresponds to the double layer volume and when this is small the charge effects will be concentrated.

(*c*)   Dye ions will be taken up without hindrance by a substrate of opposite charge but in so doing they will reduce the potential of the substrate eventually to zero when the adsorbed ion concentration equals that of the fixed charges. Since further absorption will meet the constraints discussed above, the substrate may appear to be saturated with dye ions at that point.

Clearly the whole question of coulombic interactions is of considerable importance in relation to dye sorption. However it should be remembered that the factors involved depend upon charge and potential effects alone. No distinction is made between ions of the same charge although these might be as simple as chloride or as complex as benzene azo 2 naphthol 5 sulphonic acid. Since such ions clearly differ markedly in their behaviour other factors are of great importance and with molecules as complex as dyes these generally predominate.

F   Electrokinetic effects

Due to the forces between the charged surface and the ions in solution, when the liquid is caused to stream past the surface there is a tendency to separate to some extent the counter-ions and co-ions. This establishes an electrokinetic or zeta potential. The value of the potential is that at the plane of shear of the solution near the surface and it can have many effects. The establishment of a potential by the work done to cause the liquid to stream is reflected as a drag effect resisting flow. Thus any tendency to develop a zeta potential will make it more difficult to cause liquid flow. This effect can be seen particularly in such processes as filtration under pressure. Davies and Rideal report reductions of as much as 50% due to this cause in flow rates through filters.

The reverse effect can also operate. If a potential is applied to the system then this will cause either the solution to flow (electroosmosis) or the "surface" to move (electrophoresis). Both are used of course in analytical studies and are familiar.

The electrokinetic potential being established at a little distance from the surface will be less than the potential, ψ_0, at the surface and would be expected on general grounds to approach the Donnan potential. S. M. Neale has carried out calculations of the zeta potential in the case of plugs of cotton fibres based on measurements of the streaming current. This is the current which must flow as a consequence of the existence of the potential. The values he obtained were very close to those predicted by the Gouy equation. Very little experimental work has been carried out with fibrous materials due to problems of reproducibility and consequently few conclusions can be drawn with any degree of certainty.

III DIPOLE INTERACTIONS AND PERTURBATIONAL FORCES

A General

The development of an equation of state for real gases by van der Waals involved the consideration of the real volume of real molecules and the attractive forces between molecules. This led to the equation

$$\left[p + \left(\frac{a}{v^2} \right) \right] (v - b) = RT \tag{1.56}$$

in which the intermolecular forces lead to the constant a. This reflected in van der Waals' view a cohesive or internal pressure. The effect is particularly important in cohesive states such as liquids or solids. In a liquid in particular the internal pressure is the resultant of the forces of attraction and repulsion. In some cases they can be very high, e.g. ca 2×10^4 atmospheres for water at $25°C$. These forces operate in non-ionized systems and clearly derive from forces other than the simple coulombic interactions considered already. They operate between molecules which have a dipole moment but also between molecules with no such internal charge separation, e.g. inert gases. Early attempts to explain van der Waals' forces were based on interaction between permanent or induced dipoles. The latter consideration due to Debye was the first to introduce perturbational concepts since it considered the dipole interaction between a permanent dipole in one molecule and the dipole induced in a second molecule as the two approached one another.

The dipole model fails to account for attractive forces between symmetrical molecules, e.g. Inert gases, and it was not until 1930 when London applied quantum mechanical analysis to the problem that a general explanation became available. The London approach considered the implications of the Heisenberg uncertainty principle that all molecules must possess energy even in their lowest energy states. Consequently even in the simplest molecules the nuclei and the electrons must be vibrating relative to one another. This means that at any instant in time there exists in any molecule a dipole. Although over even a very short period of time, the relative motions of the nuclei and the electrons will average out to give a resultant dipole moment of zero, the molecule will create a fluctuating field which will follow the oscillating effect of the instantaneous dipoles. Thus two molecules approaching one another may induce in one another effects similar to getting into phase so that separation of the molecules requires the input of energy and an intermolecular force exists.

Thus dipole interactions may be seen to be quite general arising from different effects but in a sense governed by similar rules. The interaction potential energy U between two permanent dipoles of dipole d at a distance r is given by

$$U = -\frac{2}{3}\left(\frac{d^4}{r^6 kT}\right) \tag{1.57}$$

where a permanent dipole interacts with an induced dipole the expression becomes

$$U = -\left(\frac{2\alpha d^2}{r^6 kT}\right) \tag{1.58}$$

in which α is the polarizability of the molecule in which the dipole is induced. The polarizability is defined as the ratio of the electrical moment of the induced dipole and the field strength operating on the molecule.

When attractive force arises from a mutual induction effect the potential energy according to London becomes

$$U = -\frac{3}{2}\left|\frac{I_a I_b}{(I_a + I_b)}\right|\frac{\alpha_a \beta_b}{r^6} \tag{1.59}$$

in which I_a and I_b are the ionization potentials of the two molecules. Since this relates to the energy of transition to a higher energy state, London took the view that the ionization potential could be replaced by hv_0 where h is Plank's constant and v_0 is the characteristic frequency. For simple substances I and hv_0 are almost identical while with more complex substances such as dyes these are characteristic transitions which are not ionizations and the term hv_0 has the more general application. It is readily assessible from the electronic energy absorption curve. The use of the term v_0 provides another link with electronic absorption. The refractive index of a substance is given by

$$n = 1 + \sum \left(\frac{a}{v_0 - v^2} \right) \tag{1.60}$$

at a frequency v. a is constant. The link between this kind of binding force and optical dispersion explains the use of the term *dispersion forces* to describe them. It will be noticed that all cases the potential energy of intermolecular attraction arising from dipole effects display an r^{-6} dependence.

Since the introduction of London's ideas further theoretical developments have taken place with a view to dealing with real situations which do not conform with those implicit in the original argument. For example dipole–quadrupole and quadrupole–quadrupole interactions have been taken into account bringing in a terms in r^{-7} and r^{-8}. The terms in r^{-6} and r^{-8} in the dispersion force energy arise from the fluctuating dipole interaction which appears in the derivation as $(r^{-3})^2 = r^{-6}$ and the fluctuating quadrupole interaction as $(r^{-4})^2 = r^{-8}$ terms respectively. Where the molecules are non-symmetrical then there is a coupling of the dipole and quadrupole terms to give a term in $(r^{-3})(r^{-4}) = r^{-7}$. The full derivation of this term involves the consideration of polarizability of the molecules along different molecular axes and is related to the optical rotatory power of the molecules. The possible relevance of this term is suggested, in the dyeing field, by the work of Bradley, Brindley and Easty on the differential adsorption of mandelic acids by wool. Some consideration of the interaction of molecules with surfaces has also resulted from Derjaguin and his colleagues.

The consideration of the question of dispersion forces this far has taken into account only two component systems. The consideration of three component systems, e.g. the interaction of solute molecules with one another or with a surface, has received recent attention. The treatment which may be used in these circumstances has been called *susceptibility theory* because of the need to take into account the rate of response of the molecule to the electrical field in which it finds itself. The difference in the polarizabilities of the solvent and solute molecules are also important in this treatment. Susceptibility theory is very effective in dealing with a wide variety of situations.

The adsorption process, of course, involves the approach of a complex molecule to an assemblage of molecules, i.e. a surface. Under these conditions there is to be expected that there should be some degree of additivity of the separate effects of all the molecules. Casimir has calculated the effect of cooperative effects of this kind for a number of idealized situations and the distance dependence is found to be as r^{-3} as compared with r^{-6} for simple systems. It has been calculated that a measurable force will exist between two metal plates at a distance of 10^{-6} in. However to measure such a force presents great difficulties due to the problem of defining the operational distance and the long time involved in establishing an equilibrium. With small molecules which present other difficulties of measurement, equilibration times are expected to be short. The mainpoint of interest is the considerable lessening of distance dependence by the cooperative factors. To this may be added the role of solvent molecules when the interactions are taking place in a medium. This has been considered recently by McLachlan who has shown that solvent molecules should transmit dispersion forces over extensive distances by virtue of the fact that their susceptibility is different from that of the solute molecules, thus enhancing the dispersion forces of attraction.

B Hydrogen bonds

A very important and widespread example of perturbational interaction with its own special characteristics is the hydrogen bond. This is much quoted in discussions of dye binding by substrates and warrants individual consideration. When a covalently bound hydrogen atom exists between two electronegative atoms, a *hydrogen bond* may be formed i.e.

Since appropriate groups, e.g. OH, NHR, CF_3 etc., are fairly common in dye molecules and substrates it is not surprising that hydrogen bonding should be postulated as a potential binding force for dyes in aggregates and with substrates. The hydrogen bond is a relatively weak one, the bond energy lying in the range $2-10$ kcal/mole. Its formation involves also a low activation energy. The strength of a hydrogen bond depends upon the electronegativity of the bound atoms so that fluorine forms much stronger hydrogen bonds than chlorine. The hydrogen bonding power of an atom can be enhanced by inductive effects and substitution so that RNH_3^+ contains a more powerful hydrogen bonding atom than RNH_2. Phenol forms more powerful hydrogen bonds through its oxygen atom than do aliphatic alcohols because of the electron withdrawing power of the phenyl ring.

When we consider a dye molecule and substrate in an aqueous medium, then if there is any possibility of the two becoming associated through a hydrogen bond then the relevant residues will be hydrogen bonded with water prior to the anticipated association. Hydrogen bonding between the dye and substrate will then require breaking the water hydrogen bonds. Alternatively hydrogen bonding groups in the substrate may be associated with one another (intramolecular hydrogen bonding) so that involvement with the dye molecule will again require two hydrogen bonds to be broken. Thus it must be presupposed that hydrogen bonding between dye and substrate will, in an aqueous system, require two hydrogen bonds to be broken for each one formed in the adsorption. Where the two broken hydrogen bonds involve water it is possible for the two water molecules freed to bond together and with an intramolecular hydrogen bond for the free group to bond with water so that there is no net change in the number of bonds. The position is shown schematically below

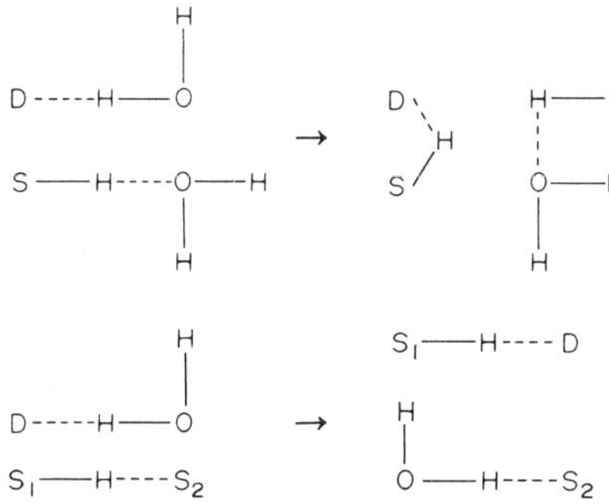

(a)

(b)

$D = $ dye
$S = $ substrate

It is difficult to envisage an energetic situation which will positively favour either of the reactions (a) and (b), and since water is normally present in an overwhelming preponderance the probability of such reactions is likely to be low. Should other forces operate in such a way that the formation of a hydrogen bond becomes probable then the hydrogen bonding energy will contribute to the stability of the dye-substrate complex. This kind of co-operative mechanism is a distinct possibility in many cases and may be regarded as playing an important role in some cases to be considered later.

Another form of hydrogen bonding is one in which the delocalized electrons of a conjugated ring system provide an electronegative centre for involvement in a bond. For example benzene will interact with proton donor substances, e.g. methanol, to form a hydrogen bond.

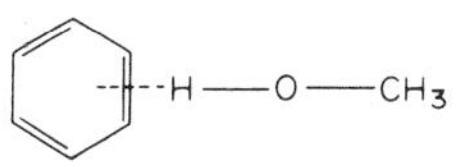

Substituting the benzene or extending the conjugated system increases the strength of such interactions. The same competitive factors will operate relative to hydrogen bonding of this kind as in the normal case, but clearly a dye molecule may offer considerable hydrogen-bonding potential through several aspects of its structure and this will tend to increase the probability of H-bonding. Additionally the effect would cause the displacement of several molecules of bound water from the surface with a consequent entropy gain. The kind of interaction discussed here may be termed a $\pi - H$ bond.

The idea that hydrogen bonding is an important contributor to dye-substrate interactions in aqueous systems arose from an observed correlation between the characteristic distance between groups in the crystalline region of cellulose and that between hydrogen bonding centres in certain dyes with good affinity for cellulose. However dyes penetrate only random regions of the polymer where there is no fixed repeat distance and many dyes have good affinity without the "necessary" spatial characteristics.

Giles has examined the possibility of hydrogen bonding between dyes and soluble polysaccharides without producing any evidence of such interactions.

The question of hydrogen bonding between dyes and substrates in the presence of water is complicated. It is very possible that it occurs in certain cases but it is probable that there will be many cases where it makes no contribution to dye binding. In non-aqueous systems, e.g. dyeing from the vapour phase or from dry solvents, hydrogen bonding is very feasible. In addition hydrogen bonding is to be expected in the dried dyeing.

IV   HYDROPHOBIC INTERACTIONS AND THE ROLE OF WATER STRUCTURE

So far binding forces have been discussed in terms of the direct attraction (or repulsion) of the molecules or groups involved. Some consideration has been given to the possible competitive role of water in relation to hydrogen bonding and also the likely advantages of having a dyebath solvent of different polarizability from that of the dye. However water can also play more important roles in relation to binding forces provided a considerable proportion of the free energy created by the binding process. These arise from the unusual properties and structure of water as a material.

The existence of a high degree of internal bonding in liquid water is evidenced by its high boiling point. As hydrogen oxide water would be expected to boil at $-161°C$ instead of $+100°C$. It has other interesting properties which may be gleaned by a comparison of ice and water.

	Liquid water	Ice
$[H^+]$	1×10^{-7} molar	1×10^{-10} molar
Conductance	1×10^{-8} ohm^{-1} cm^{-1}	3×10^{-9} ohm^{-1} cm^{-1}
Mobility of H^+	$3\cdot6 \times 10^{-3}$ cm^2 volt^{-1} sec^{-1}	$3\cdot0 \times 10^{-1}$ cm^2 volt^{-1} sec^{-1}
Mobility of Li^+	4×10^{-4} cm^2 volt^{-1} sec^{-1}	$<10^{-8}$ cm^2 volt^{-1} sec^{-1}
Mobility of H_2O	$2\cdot45 \times 10^{-5}$ cm^2 sec^{-1}	8×10^{-11} cm^2 sec^{-1}
k (for $H^+ + OH^- \rightarrow H_2O$)	10^{11} l. mol. sec^{-1}	$10^{13} - 10^{14}$ l. mol. sec^{-1}

A number of unusual features can be seen. Firstly $[H^+]$ ice/$[H^+]$ water equals 10^{-3} but the corresponding ratio of conductances is 0·3. Since the current is carried by the ions present the mobility of protons in ice must be of the order of 85 times greater in ice than in liquid water. Similarly the combination of protons and hydroxyl ions takes place even more rapidly in ice than in water. These observations contrast markedly with the lower mobility of lithium ions (comparable in size with protons) and the mobility of water molecules themselves.

The explanation of the anomalies arises from the fact that whereas for example lithium ions must transport themselves physically in the system, protons may move by a flow of energy through the electrical circuit provided by the extensive hydrogen bonding between water molecules. Thus

---- Intermolecular hydrogen bonds

Thus water is a highly structured material. The nature of the structure is still the subject of debate. Earlier theories suggested that water consisted of "flickering clusters" of molecules so called because of their transitory existence. Water was regarded as consisting at any one instant of molecules participating in a highly ordered (almost ice like) cluster and free molecules. The role of any particular water molecule changed with time continuously. This view has tended to be displaced by a model which presupposes a general order in which defects exist. The defects are regarded as being mobile and present to an extent depending upon temperature. The structure of pure water is relevant to the present topic because the inclusion in water of an alien solute ion or molecule or a surface will have short and longer range effects on the water structure and consequently on adsorption from aqueous solution. Two effects will be considered.

A Ionic interactions in aqueous solution

The presence of a permanent charge centre whether it be a dissolved ion or a point charge in a surface creates an electric field capable of polarizing adjacent water molecules. The nature of the polarization will clearly depend upon the sign of the charge but clearly it will contribute to the kind of local order of the water molecules. Also operating will be effects due to the physical size of the ion or the shape of the surface and also dispersion force interactions between the ion and the water molecules. There will also be the general structuring influence arising from the properties of water before the ion or charged surface arose as a factor. The combination of these forces can take two forms leading to regular and negative hydration effects. It has been shown that sodium ions are surrounded by a closely packed zone of water molecules so that the ion is positively hydrated. Chloride ions on the other hand show a negative hydration in that they are surrounded by a water molecular environment less dense than normal water. In the case of sodium ions they are dissolved in water with a fall in bulk density due to the packing effect so that the apparent specific volume is negative. The close-packing effect of certain ions operates in competition with the normal structuring effects so that there is an equilibrium condition of less than normal order. The mobility of water molecules thus falls as they approach the ion (or the charged surface) and this of course relates to the zeta potential effect. The ordering of the water molecules is important because of the loss of entropy which it represents.

If ions of opposite charge are present with hydration shells polarized in opposite senses, then electrostatic neutralization will result in the release of the polarized water. Thus the ionization of a weak acid, e.g.

$$CH_3COO^- + H^+ \rightleftharpoons CH_3COOH$$

is not represented precisely by the normal equation when the acid is dissolved in water. The reaction of the two ions result in a randomization of the water molecular present while the dissociation has the opposite effect. Thus the equilibrium would be more correctly written as

$$CH_3COO^- + H^+ + (water)_{1,t} \rightleftharpoons CH_3COOH + (water)_{2,t}$$

where the suffices 1, and 2 represent water in the two states at temperature t. The free energy contribution to an ionic interaction due to the entropy gain can be quite significant. The formation of an ion pair e.g.

$$-(NH_3^+)_{aq} + -(COO^-)_{aq} \rightleftharpoons -(NH_3^+)(COO^-)-$$

will result in a gain of 30–40 entropy units which at 60°C provides a free energy of 10–13 kcal/mole.

The anomalous effects of water structure in association/dissociation reactions are of general importance and have been extensively studied particularly in relation to weak acid/base equilibria.

B. Non-ionic interactions in aqueous solution

The dissolution of a non-polar molecule in water, while causing no polarization, causes a disturbance in the structure and creates the requirement that the structuring forces in the water rearrange themselves in the lowest energy configuration among themselves and in relation to the new alien molecules. The dissolution of non-polar molecules in water causes *an increase in heat capacity* indicating less ready thermal motion of water molecules and a lessening of the randomness of the system (loss of entropy). The ordering effect of non-polar residues can be very extensive in some cases giving crystal hydrates containing as much as 90% water which are remarkably stable e.g.

tetrabutylammonium hydroxide crystal hydrate

TBAH: water = 1·32

melting pt. 30°C

A wide range of non-polar solutes form crystal hydrates, e.g. argon, xenon, methane, chloroform, chlorine. All these disparate solutes show the same hydration energy ($16 \pm$ kcal/mole) indicating that it is the interaction of *water molecules with one another* rather than with the solute which is involved.

Non-polar solutes thus cause "structuring" of the water. Structure is not a permanent feature. What is meant is that if the mean reorientation time of a water molecule is 10^{-11} sec and a non-polar solute increases this to 10^{-10} sec then "structuring" has taken place. The more ordered region of water around a non-polar solute has been likened to an iceberg because of the greater ice like character of the water in it. This description can be deceptive if taken too literally, but may be used with caution for the purposes of discussion of how the "icebergs" occur.

Five energy states for liquid water molecules may be postulated depending upon the extent to which hydrogen bonding is developed. In addition to the creation of free energy by hydrogen bond formation there will also be a dipole-dipole energy between near neighbours. Opposing the bonding which will result from these driving forces will be thermal energy so that the average condition will depend on temperature.

A water molecule not involved in hydrogen bonding can accommodate eight near neighbouring water molecules while a water molecule involved in the maximum number of four hydrogen bonds can accommodate only four near neighbours. Intermediate levels of hydrogen bonding imply different possibilities with regard to unbonded near neighbours.

If a tetra hydrogen bonded water molecule is regarded as being in the ground energy state, then other degrees of hydrogen bonding will result in water molecules occupying higher energy levels each separated from the one lower down by $0.5\,E_h$. This value is made up by the difference between free energy lost due to hydrogen bond rupture and that gained from the new neighbour interactions. The energy is halved because two molecules are involved in the process. The latter is shown schematically overleaf.

When a non-polar solute is present then some water molecules find that as a near neighbour there is a non-polar solute molecule instead of water

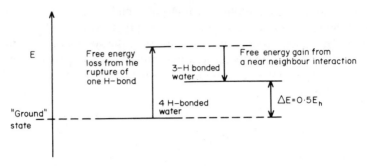

molecules. Since the dipole–dipole interaction energy between two water neighbours exceeds the dispersion force energy between the water molecules and the non-polar solute molecule, the average energy level of all incompletely hydrogen bonded water molecules will rise by Er which measures the average net difference. However a completely hydrogen bonded water molecule can accept a non-polar near neighbour without affecting other interactions. This is only possible due to the electrical inertness of the solute allowing only dispersion force interactions. The new fifth neighbour becomes involved in a cage structure as a consequence.

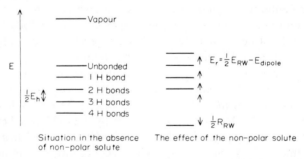

4 H–bonded water   Interaction with a fifth neighbour (⊗)

----- Signifies interactions rather than bonds

The new interaction results in a lowering of the energy level by an amount which equals the interaction energy. Schematically the situation may be shown as follows.

The effect is thus to favour multiple hydrogen bonding in the neighbourhood of the non-polar solute molecule with the development of structure on a localized basis.

Dissolution of the non-polar solute thus depends upon the existence of structure and consequently the solubility will fall with increasing temperature. Dissolution will be exothermic (hydrogen bonding) and accompanied by a loss of entropy. The precipitation process, as the temperature is raised, results from a drive to regain entropy. In other words the non-polar solute molecules are driven to interact with one another by an entropic force within the water. This is known as a *hydrophobic interaction* and operates as a bonding force between non-polar solute molecules as well as non-polar regions or residues of large molecules, e.g. dyes and polymers.

V FURTHER READING

Adamson, A. W., *The Physical Chemistry of Surfaces*, Interscience (1960).
Bell, R. P., *The Proton in Chemistry*, Methuen (1959).
Covington and Jones (ed.), *Hydrogen-bonded Solvent Systems*, Taylor-Francis (1968).
Davies, J. T. and Rideal, E. K., *Interfacial Phenomena*, 2nd edition, Academic Press (1963).
Discussions of the Faraday Society No. 16: *The Physical Chemistry of Dyeing and Tanning*. (1954).
Discussions of the Faraday Society No. 40: *Intermolecular forces*. (1966).
Eigen, M., *Proton Transfer*, Augew. Chem. Internat. Ed. 3 (1964) 1.
Haydon, D. A., *Recent Advances in Surface Chemistry*, Vol. 1, (1965).
Neurath, H. (ed.), *The Protons*, 2nd edition, Vol. 1, Chapter 6. Academic Press (1963).
Pontificae Academiae Scientiarum Scripta Varia No. 31: *Molecular Forces*, John Wiley & Sons Inc., New York (1967).
Samoilov, O. Ya., *The structure of aqueous electrolyte solutions*, Consultants Bureau.

Chapter 2

The Adsorption of Dyes
and Adsorption Equilibria

I   INTRODUCTION

The adsorption of dyes involves, according to circumstances, dyes which may be charged or not, interacting with substrates which may be oppositely or similarly charged or not charged at all. The substrates may be swollen or not and may have different stabilities in relation to the dyeing process. Consequently the wide variability in the context of the dyeing process makes it impossible to develop a completely general dyeing theory except in relation to a hypothetical substrate. However there is enough common ground between the particular situations which have to be considered to permit a limited generalized viewpoint to be developed. This can be useful when taken into account with the special factors relating to special cases.

The simplest adsorption situation is a *two-component system*, adsorbate and adsorbant, in which the adsorbate is taken up from the gas phase. Since many non-polar dyes possess a significant vapour pressure this type of system has been studied in relation to dyeing, and has been the basis for the most extensive studies generally. The consideration of only two components allows the theory of gases to be applied and activity factors simplified. From a consideration of two components systems, we shall go on to consider the particular *three-component system* in which dye and water molecules are in the gaseous phase. Finally the three components *solute–solvent–sorbate* system will be discussed.

II SIMPLE TWO-COMPONENT SYSTEMS

If a number of gas molecules are kept within a closed system they will exert a measurable pressure on the walls of the container due to collision with it. The pressure will depend upon the number of collisions per second and the kinetic energy of the molecules. Since the molecules are real they will not recoil from the surface in zero time. In fact the molecules will stay for at least the period of a molecular relaxation (10^{-13}–10^{-14} seconds). Consequently there will always be a surface concentration equal to $n\tau_0$ in which n is the number of collisions per second and τ_0 is the period of a molecular relaxation. This assumes that there is no interaction between the impinging molecules and the surface. If such a reaction does occur then the dwell time will become τ ($>\tau_0$) and there will be a surface excess of gas molecules. These are the adsorbed molecules and the quantity $n\tau$ is fundamental to the study of adsorption.

The value of n may be reached by application of the kinetic theory of gases. For an ideal gas n is given by

$$n = \frac{Np}{\sqrt{(2\pi MRT)}} \tag{2.1}$$

in which R is the gas constant, N is Avogadros's number, p is the pressure, M is the molecular weight and T is the absolute temperature. Expressing the pressure p in millimetres of mercury and combining the constants

$$n = 3.52 \times 10^{22} \times p\sqrt{(MT)^{-1}} \tag{2.2}$$

Consequently n is a very large number except at very high temperatures and very low pressures. For example for oxygen gas at 20°C and normal pressure $n = 2.75 \times 10^{23}$. Even at very low pressures, e.g. $p = 1 \times 10^{-5}$ mm of mercury at 20°C n for oxygen is 3.63×10^{15}. This means that ideally equilibria in two component systems should establish themselves very rapidly since the available surface area of typical absorbants is such that about 10^{15} molecules will give complete coverage. In fact matters are not so simple since other factors have to be considered.

One of the factors limiting the rate of collision with the surface and hence the rate of equilibration is the collision of gas molecules with one another. This limits the *mean free path* of the molecules and hence the rate at which they travel. The mean free path is given (approximately) by the equation,

$$L \simeq 10^{-5} \times \frac{T}{273} \times \frac{760}{p} \, \text{cm} \qquad (2.3)$$

in which p is expressed in millimetres of mercury. It can be seen that not until p approaches 10^{-5} mm does L become at all large.

One point which must be remembered is that despite the achievement of an adsorption equilibrium in which the number of molecules arriving at the surface is equal to the number departing from it, the rate of adsorption equals the rate of desorption, the two processes have nothing to do with one another. The rate of adsorption is related only to the properties of the gas, i.e. p, M and T. The rate of desorption is related to the nature of the interaction between adsorbed gas molecules and the surface. A good example is provided by a vapour in equilibrium with its own liquid. The rate of evaporation is related to temperature and latent heat while the rate of condensation is related to the basic gas factors, p, M and T.

If the mean free path were very large and all molecules escaping from a liquid were prevented in some way from recondensing then rates of evaporation would be very high indeed, e.g. the sea would evaporate at a rate of 9 m depth per hour instead of the 0·5–2 m per annum observed in practice.

The quantity τ has been measured both directly and indirectly in a number of cases at different temperatures. It is found that the values fit an equation put forward by Frenkel

$$\Upsilon_0 \tau = \tau_0 \exp Q/RT \qquad (2.4)$$

in which τ_0 is the period of the molecular vibration of an adsorbed molecule perpendicular to the surface, and Q is the heat of adsorption. Experimental data applied to this equation are always found to give a value of τ_0 of the order of $10^{-13} - 10^{-14}$ as expected. The quantity Q, the heat of adsorption, is the most important factor consequently in determining the value of τ, the time of absorption or dwell time.

Applying Frenkel's equation and giving Q an arbitrary value of 1·5 kcal/mole $\exp Q/RT$ is about 10 at 20°C. When Q reaches values of the order of 15 kcal/mole τ reaches about 2×10^{-2} seconds which is a very considerable increase in dwell time over τ_0.

If we now calculate values of $n\tau$ to give the surface concentration very considerable values greatly in excess of that possible with known surface areas are reached with quite low values of Q. This is due to the fact that it has been assumed up to now that every collision with the surface is successful, i.e. leads to adsorption. This is clearly incorrect for most surfaces since the occupation of part of the surface must reduce the probability of a successful collision with it. In addition since the gas molecules will not be symmetrical except in unusual circumstances, the arriving molecule will not always be disposed favourably for the adsorption interaction to occur. In addition the arriving molecules must have the right energy for the adsorption. Consequently the adsorption rate and the equilibrium situation will be governed by a sticking probability P which will be given by

$$P = \sigma.f(\theta)\exp(-Q/RT) \tag{2.5}$$

in which σ is the entropy term governing the disposition of the molecules, $f(\theta)$ is some function of the degree of surface occupation, and Q is the activation energy of adsorption. Since the term for the sticking probability takes the same form as Frenkel's equation it is clearly related to the dwell time. The rate of adsorption u_a will be the product of the number of collisions and the sticking probability, i.e.

$$u_a = \frac{Np}{\sqrt{(2\pi MRT)}}.P = \frac{Np\sigma f(\theta)}{\sqrt{(2\pi MRT)}}\exp(-Q/RT) \tag{2.6}$$

Since allowance must be made for the possibility that the adsorbing surface is not homogeneous and may contain adsorption sites of different properties,

$$u_a = \frac{Np}{\sqrt{(2\pi MRT)}}\sum_{i=1}^{i=i}\sigma_i f(\theta)_i \exp(-Q_i/RT) \tag{2.7}$$

This equation is not soluble without a very great deal of additional information. General solutions can be found, however, if the surface is considered to be *uniformly heterogenous*. This means that any small element of the surface, *ds* contains all the elements of the total surface in the same distribution.

This assumption enables us to integrate the expression based on the incremental area between the limits θ and 0, i.e. between a fraction θ of the surface and zero coverage. Thus

$$u_a = \frac{Np}{(2\pi MRT)}\int_0^\theta \sigma_s.f(\theta)_s \exp(-Q_s/RT)\,ds \tag{2.8}$$

In order to proceed assumptions have to be made about the relationship between σ_s, $f(\theta)_s$, Q_s and ds. For a given interaction σ_s may be regarded as constant and a function more of the absorbate than the absorbant. Also θ being the fractional surface occupation, $f(\theta)$ may be put equal to $(1-\theta)$ and equal to the probability of collision with an unoccupied but adsorbing part of the surface. Hence

$$u_a = \frac{Np\sigma(1-\theta)}{\sqrt{(2\pi MRT)}}\int_0^\theta \exp(-Q/RT)\,ds \tag{2.9}$$

It will be seen later that the further development of theoretical models depends upon the next assumption regarding the relationship between Q and S (or θ), i.e. the nature of the surface heterogeneity.

For example Q may be assumed to be proportional to θ so that

$$Q = Q^o - \alpha\theta \tag{2.10}$$

where α is a constant and Q^o is the interaction energy with the most active absorption site. The equation then becomes

$$u_a = \frac{Np\sigma(1-\theta)}{\sqrt{(2\pi MRT)}} \int \exp - \frac{(Q^o - \alpha\theta)ds}{RT} \tag{2.11}$$

$$\text{Since } \exp - \frac{(Q^o - \alpha\theta)}{RT} = \exp\left(\frac{\alpha\theta}{RT}\right)\exp\left(\frac{-Q^o}{RT}\right) \tag{2.12}$$

$$u_a = \frac{Np\sigma(1-\theta)\exp(-Q^o/RT)}{\sqrt{(2\pi MRT)}} \left| \exp\frac{\alpha\theta}{RT} \right|_0^\theta \tag{2.13}$$

$$= A \exp \alpha\theta/RT$$

where A is a constant. This is so if θ does not approach 0 or 1 so that the main variable is the exponential term.

This form of relationship between the rate of uptake and the degree of surface occupation is observed experimentally in numerous cases and is incorporated in the *Elovich equation*

$$dq/dt = a \exp(-bq) \tag{2.14}$$

in which q is the quantity adsorbed and a and b are constants. It must be remembered that correspondence between the Elovich equation and the theoretical rate equation does not permit fundamental significance to be given to the constants a and b. The theoretical equation merely expresses what will be the effect if certain assumptions are made. That the prediction corresponds with experiment does not justify the assumptions unless other possible assumptions have been shown to be inapplicable. At the same time the correspondence between theory and practice does suggest the kind of variable which might give rise to the Elovich equation being applicable.

A   Adsorption isotherms, Isobars and Isosteres

The question of adsorption kinetics which has been considered so far is rather more complex where textile and other very complex surfaces are concerned and merits a special treatment of its own. For the present the equilibrium state will be considered, i.e. the state in which adsorption and desorption states are equal. What interests us is the relationship between the amount of adsorbate taken up and the amount in the external phase. Experimental data may be expressed in various ways.

Isotherms   the relationship between the concentration of adsorbate in the two phases *at a given temperature.* If we consider an ideal gas which obeys the equations previously considered, the concentration on the surface, Γ, is given by

$$\Gamma = n\tau = \left[\frac{Np}{\sqrt{(2\pi MRT)}}\right] . \tau_0 \exp(Q/RT) \tag{2.15}$$

so that given a particular surface, a particular adsorbate,

$$\Gamma = \left(\frac{kp}{\sqrt{T}}\right) \exp(Q/RT) \tag{2.16}$$

Consequently at a given temperature, if Q is constant

$$\Gamma = k'p \tag{2.17}$$

where

$$k' = \left[\frac{N}{\sqrt{(2\pi MRT)}}\right] \exp(Q/RT) \tag{2.18}$$

This is the simplest form of an isotherm equation. Other more complex relationships arise when probability factors are functions of Γ, adsorbate molecules interact, heat of adsorption is not constant etc.

Isobars   the relationship between the concentration of adsorbate in the surface and the temperature when the external concentration of adsorbate is constant. This is equivalent to the situation of equation (2.16) where p is constant and T is a variable, i.e.

$$\Gamma = k'' \frac{\exp(Q/RT)}{\sqrt{T}} \tag{2.19}$$

where

$$k'' = \frac{Np}{\sqrt{(2\pi MR)}} \tag{2.20}$$

Due to the greater sensitivity of Γ to the exponential term as compared with $\sqrt{T}$, Γ tends to fall exponentially with T where the gas is ideal. It is sometimes observed in isobars that at certain temperatures there is a sudden marked change in Γ over a narrow temperature range. This normally indicates a change in the mechanism of adsorption with temperature

Isosteres   the relationship between the concentration of adsorbate in the external phase and the temperature when the surface concentration is constant i.e.

$$p = k''' \sqrt{T} \exp\left(-Q/RT\right) \qquad (2.21)$$

where

$$k''' \frac{[\Gamma \sqrt{(2\pi M R)}]}{N} \qquad (2.22)$$

Since equation (2.21) is much more sensitive to the exponential term than it is to $\sqrt{T}$, the value of Q may be calculated from the relationship between ln p and $1/T$.

All three forms of equilibrium relationship are employed in adsorption studies. The most commonly used is the *isotherm*. Experimental isotherms obey equation (2.17) in only a minority of cases. Normally there are departures from ideality which may be described (but not necessarily explained) by empirical equations or equations based on theoretical models. Factors giving use to non-ideal behaviour include

(1) modification of the adsorbant by the adsorbate
(2) heterogeneity of the adsorbing surface
(3) variation of the probability of adsorption with Γ
(4) interaction of adsorbate molecules on the surface or in the external phase.

Experimental isotherms generally correspond with one of five types.
The isotherms are discussed in more detail below.

All of the isotherms may be derived theoretically on the basis of postulated models of the adsorption process. However it must always be remembered that if a set of experimental data conform with one of the classical isotherms this does not constitute any proof that the data fit the theoretical model situation. In almost all cases different models can be postulated to give the same predicted relationship.

1. *The partition isotherm*

This is predicted from a consideration of ideal gas theory (c.f. equation 2.16) but its observation does not mean ideal adsorption in the complete sense. This may be explained more clearly in relation to dissolved adsorbate and will be considered in relation to three component systems.

2. *The Langmuir isotherm*

This was derived by Langmuir on the basis of treating the adsorption and desorption processes as chemical reactions reversible in character. The isotherm equation is derived for a surface containing a finite number of identical adsorption sites and an adsorption process such that occupation of any site does not effect the characteristics of any other site. Thus

$$\text{the rate of adsorption} = k_a p \left(S - \Gamma\right) \qquad (2.23)$$

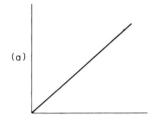

(a)

Partition isotherm
(corresponds to the ideal gas
isotherm)

$$\Gamma = kp$$

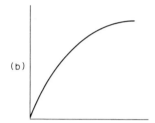

(b)

Langmuir isotherm

$$1/\Gamma = k_1 + k_2/p$$

(c)

Freundlich isotherm

$$\Gamma = k_1 p^{k_2} \text{ where } k_2 < 1$$

(d)

Temkin isotherm

$$|\Gamma = k_1 \log p + k_2|_{\Gamma \sim 0.5}$$

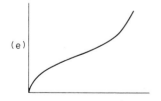

(e)

*Brunauer, Emmett and Teller
(BET) isotherm*

This isotherm may take several
forms only one of which is shown.

in which k_a is the velocity constant of adsorption, p is the external pressure and $S-\Gamma$ is the surface concentration of unoccupied sites, i.e. the difference between the total site concentration (S) and the adsorbed surface concentration (Γ).

$$\text{the rate of adsorption} = k_d\Gamma \tag{2.24}$$

in which k_d is the velocity constant of desorption.

At equilibrium the rates of absorption and desorption are equal so that

$$k_a p_e\,(S-\Gamma_e) = k_d\Gamma_e \tag{2.25}$$

where the subscript e signifies an equilibrium value. From equation (2.25)

$$\Gamma_e = \frac{k_a p_e S}{k_d + k_a p_e} \tag{2.26}$$

or

$$\frac{1}{\Gamma_e} = \frac{k_d}{k_a.S}\cdot\frac{1}{p_e}+\frac{1}{S} \tag{2.27}$$

which conforms to the empirical equation describing the isotherm and gives some significance to the experimental constants k_1 and k_2.

3. The Freundlich and Temkin isotherms

The Freundlich isotherm is widely observed in practice and its mathematical expression

$$\Gamma = k_1 p^{k_2} \qquad (k_2 < 1) \tag{2.28}$$

was derived quite empirically. Equations of the form of equation (2.28) can be derived mathematically by assuming that the adsorbing surface is uniformly heterogeneous and sites of the jth type will be fractionally occupied in accordance with their own Langmuir equation so that

$$\frac{\Gamma_j}{S_j} = \theta_j = \frac{k_j p}{1 + k_j p} \tag{2.29}$$

If the distribution of different kinds of site is assumed to be exponential and that the adsorption constants similarly vary so that

$$n_j = n_o \exp(Va) \tag{2.30}$$

$$k_j = k_o \exp(Vb) \tag{2.31}$$

in which sites of type o are the last to be occupied (i.e. they are the least active); k_o is the adsorption constant of such sites; a and b are constants; V is the variable factor or that property which differs between the sites causing the heterogeneous character of the surface, e.g. activation energy of sorption, entropy of sorption etc., then equations of the Freundlich type may be derived. Because such equations may be derived in so many ways, Freundlich behaviour should never be taken to indicate any particular type of surface heterogeneity although it does demonstrate that such heterogeneity exists.

If another assumption is made so that

$$V = V_o(1 - \alpha\theta) \tag{2.32}$$

i.e. the variable factor is proportional to θ, and equation (2.31) is assumed to apply, another solution is obtained which is the Temkin isotherm equation. At the present time it is not possible to speculate usefully on the probable physical differences between Freundlich and Temkin surfaces but there is little doubt that they represent differences in forms of heterogeneity.

4. The BET isotherm

Isotherms of the BET type are frequently encountered in practice and are used widely in powder technology for surface area estimations. The mathematical derivation of the isotherm is based on the assumption that monolayer adsorption onto the adsorbant surface is accompanied by the formation of multilayers of adsorbed molecules. In considering the adsorption-desorption equilibrium it is assumed that only the outermost layers of adsorbed molecules are involved. This is represented diagrammatically in Fig. 2.1.

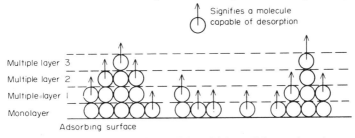

Fig. 2.1. A schematic representation of the BET model for multilayer adsorption.

If θ_1 is the fraction of the surface covered with a monolayer of molecules, θ_2 the fraction covered with a double layer etc. then the total surface concentration Γ is given by

$$\Gamma = \Gamma_0\theta_1 + 2\Gamma_0\theta_2 + 3\Gamma_0\theta_3 + \dots i\Gamma_0\theta_i \tag{2.33}$$

Thus

$$\Gamma = \Gamma_0 \sum_{i=1}^{i=\infty} i\theta_i \tag{2.34}$$

At equilibrium all the fractions are constant and this must include the fraction of surface unoccupied (θ_0) which is given by

$$\theta_0 = 1 - \theta_1 - \theta_2 - \theta_3 - \dots = 1 - \sum_{i=1}^{i=\infty} \theta_i \tag{2.35}$$

At equilibrium, the number of molecules striking the unoccupied surface must equal the fraction θ_0 of the total number n, and also the number desorbed from the monolayer, i.e.

$$n\theta_0 = v\Gamma_0\theta_1 \tag{2.36}$$

in which v is a frequency factor (related to Γ) and which is analgous to the Langmuir rate constant of desorption.

Similarly for the monolayer, there must be an equality between the rate of evaporation from the double layer plus the rate of adsorption onto free surface (forming the monolayer) and the rate of evaporation from the mono-layer plus adsorption onto it (loss of monolayer). Schematically this is represented in Fig. 2.2.

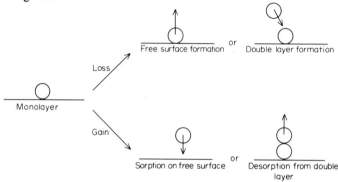

Fig. 2.2. Equilibria involving the surface monolayer in the BET model for multilayer adsorption.

Algebraically this equilibrium is expressed as

$$n\theta_0 + v_1\Gamma_0\theta_2 = v\Gamma_0\theta_1 + n\theta_1 \tag{2.37}$$

Because of equation (2.36)

$$v_1\Gamma_0\theta_2 = n\theta_1 \tag{2.38}$$

and in general

$$v_{i-1}.\Gamma_o\theta_i = n\theta_{i-1} \tag{2.39}$$

The frequency factors v_i are related to the dwell times τ_i and since the interaction of adsorbed molecules in the first and succeeding multilayer is with other adsorbed molecules it may be assumed quite reasonably that

$$\tau_1 = \tau_2 = \tau_3 \ldots = \tau_i \tag{2.40}$$

$$v_1 = v_2 = v_3 \qquad = v_i$$

The equations may now be simplified.

$$\theta_2 = \left(\frac{n}{v_1\Gamma_0}\right).\theta_1 = x.\theta_1 \tag{2.41}$$

$$\theta_3 = \left(\frac{n}{v_2\Gamma_0}\right).\theta_2 = x^2\theta_1$$

$$\theta_i = x^{i-1}\theta_1 = x^i\left(\frac{v_1\theta_0}{v}\right) \tag{2.42}$$

since
$$\theta_1 = \frac{n\theta_0}{v\Gamma_0} = x\left(\frac{v_1.\theta_0}{v}\right) \tag{2.43}$$

Substituting equation (2.42) in equation (2.34)

$$\Gamma = \Gamma_0 \sum_{i=1}^{i=\infty} i\theta_i = \frac{\Gamma_0 v_1}{v}\theta_0 \sum_{i=1}^{i=\infty} ix^i \tag{2.44}$$

Solving equations (2.35) and (2.42) simultaneously gives

$$\theta_0 = \left(1 + \frac{v_1}{v}\sum_{i=1}^{i=\infty} x^i\right)^{-1} \tag{2.45}$$

If we now call $v_1/v = k$ and substitute for θ_0 in equation (2.44)

$$\Gamma = \frac{\Gamma_0 k\sum ix^i}{1+k\sum x^i} \text{ between the limits } i = \begin{vmatrix} \infty \\ 1 \end{vmatrix} \tag{2.46}$$

The two series in x^i may be summed in the usual way giving

$$\Gamma = \frac{k\Gamma_0 x}{(1-x)^2}.\frac{1}{1+kx/(1-x)} \tag{2.47}$$

$$\Gamma = \frac{k\Gamma_0 x}{(1-x)(1-x+kx)} \tag{2.48}$$

Going back to equation (2.1) it can be seen that n is proportional to the pressure, p. Also x is dimensionless so that in

$$x = \frac{n}{v_1\Gamma_0} = \left(\frac{\beta}{v_1\Gamma_0}\right).p \tag{2.49}$$

the term $\beta/v_1\Gamma_0$ has the dimensions p^{-1} in order to keep x dimensionless and we can replace it by a conceptual pressure q which may be calculated from β, Γ_0 and v_1, i.e.

$$x = p/q \tag{2.50}$$

and substituting for x in equation (2.48) gives

$$\Gamma = \frac{kp\Gamma_0}{(q-p)(1+(k-1)p/q)} \tag{2.51}$$

More usefully the surface concentrations can be expressed in terms of volume of adsorbed gas, i.e.

Γ is proportional to the total volume adsorbed

Γ_0 is proportional to the adsorbed monolayer volume, v_m

The proportionality factor is in both cases a function of the surface area so that the usual form of the BET equation

$$v = \frac{kpv_m}{(q-p)[1+(k-1)p/q]} \tag{2.52}$$

may be used for surface area calculations.

Normally the value of q is assumed to equal the saturation pressure of the adsorbate at the experimental temperature because when p approaches q, v becomes infinite. If the maximum value of p is p_o then p_o must equal q and the equation may be rewritten,

$$\frac{p}{(p-p_o)v} = \frac{1}{v_m k} + \frac{k-1}{v_m k} \cdot \frac{p}{p_o} \tag{2.53}$$

which gives a graphical solution for v_m and k when v, p and p_o are known.

B   Simple two-component dyeing systems

Clearly simple two-component dyeing systems are experimentally accessible to a limited degree. Only dyes with significant vapour pressure may be used. Sufficient examples of such dyes are provided by the very weakly polar disperse dyes and there has been extensive study of this kind of dyeing system in recent years. The system has more than theoretical significance since in the Thermosol dyeing process dye transfers from the solid to the dyed state or from one kind of fibre to another through the vapour phase.

The vapour phase application of disperse dyes to appropriate substrates, e.g. polyester and cellulose acetate, is observed to follow the ideal partition isotherm. Jones and Seddon have shown that the isosteric heat of adsorption is independent of Γ and that it is of the same magnitude as the heat of condensation of the vapour. This shows that the binding forces between dye and substrates are such that an ideal mixing situation can exist and the dye-fibre complex behave like an ideal solution. The requirements for this to be so are calculable.

A simple consideration of the physical factors in dye adsorption by a polymer would lead to the conclusion that a "solution" of a large dye molecule in a polymeric matrix as solvent would be inevitably non-ideal. However this is not so.

If it is assumed that interactions between molecules in the "solution" are effectively confined to adjacent molecules and that each molecule of each kind (solvent and solute) is surrounded by c neighbours, then the effective energy of solute molecules with respect to the system will be

$$c\left[\frac{n_1}{n_1+n_2}\right]\Pi_{11}+c\left[\frac{n_2}{n_1+n_2}\right]\Pi_{12} \tag{2.54}$$

where Π_{11} and Π_{12} are the interaction energies of solvent molecules with one another and with the solute molecules respectively. As there are n_1 and n_2 molecules of solvent and solute molecules the total interactions with respect to each kind of molecule are

$$c\left[\frac{n_1^2}{1+n_2}\right]\Pi_{11}+c\left[\frac{n_1 n_2}{n_1+n_2}\right]\Pi_{12} \text{ for solvent} \tag{2.55}$$

$$c\left[\frac{n_1 n_2}{n_1+n_2}\right]\Pi_{12}+c\left[\frac{n_2^2}{n_1+n_2}\right]\Pi_{22} \text{ for solute} \tag{2.56}$$

The total potential energy of the system (U) as a consequence of mixing will be half of the sum of the components, (2.55) and (2.56). The sum is halved since each molecule must be counted only once. Thus

$$U=c\left[\frac{n_1^2}{n_1+n_2}\cdot\frac{\Pi_{11}}{2}+\frac{n_2^2}{n_1+n_2}\cdot\frac{\Pi_{22}}{2}+\frac{n_1 n_2}{n_1+n_2}\cdot\Pi_{12}\right] \tag{2.57}$$

If from two pairs of dissimilar molecules, two similar pairs are made, i.e.

then the gain in potential energy will be

$$2\Pi_{12}-(\Pi_{11}+\Pi_{12})=2\Delta\Pi \tag{2.58}$$

so that the change in potential energy, ΔU^0, for c pairs of dissimilar molecules is

$$\Delta U^0=c\Delta\Pi=c\left[\Pi_{12}-\frac{(\Pi_{11}+\Pi_{22})}{2}\right] \tag{2.59}$$

Multiplying through equation (2.59) by $n_1 n_2/(n_1+n_2)$ and substituting for $n_1 n_2/(n_1+n_2).\Pi_{12}.c$ in equation (2.57) gives on simplification,

$$U=n_1\left[\frac{c\Pi_{11}}{2}\right]+n_2\left[\frac{c\Pi_{22}}{2}\right]+\frac{n_1 n_2 .\Delta U^0.}{(n_1+n_2)} \tag{2.60}$$

If $n_1=0$, $U=n_2\left[\frac{c\Pi_{22}}{2}\right]$ \tag{2.61}

$$\text{If } n_2 = 0, \ U = n_1\left[\frac{c\Pi_{11}}{2}\right] \tag{2.62}$$

In each case U is half of the interaction energy of molecules of solute (equation 2.61) or solvent (equation 2.62) in the pure state. Since these are standard state quantities which may be written as U_2^0 and U_1^0 respectively.

$$U = n_1 U_1^0 + n_2 U_2^0 + \frac{n_1 n_2}{(n_1 + n_2)} \cdot \Delta U^0. \tag{2.63}$$

The partial molar potential energies for the two kinds are thus

$$\left[\frac{\partial U}{\partial n_1}\right]_{n_2} = U_1 = U_1^0 + \left[\frac{n_2}{n_1 + n_2}\right]^2 \Delta U^0 \tag{2.64}$$

$$\left[\frac{\partial U}{\partial n_2}\right]_{n_1} = U_2 = U_2^0 + \left[\frac{n_1}{n_1 + n_2}\right]^2 \Delta U^0 \tag{2.65}$$

and $U = n_1 U_1 + n_2 U_2$ (2.66)

which satisfies the Gibbs–Duhem relationship.

For an ideal mixing process ΔU^0 must equal zero. This condition is met according to equation (2.59) when

$$\Pi_{12} = \frac{(\Pi_{11} + \Pi_{22})}{2} \tag{2.67}$$

Clearly it is not necessary that $\Pi_{11} = \Pi_{22} = \Pi_{12}$ for this condition to hold but Π_{12} must always lie between Π_{11} and Π_{22}. In terms of a dyeing situation, if Π_{12} (the force between dye and fibre molecules) is an appropriate value greater than the bonding forces between dye molecules in the crystalline state and less than the bonding forces between the polymer molecules then the dyeing equilibrium could follow ideal mixing behaviour and a partition isotherm even though it was quite clear that dye molecules and polymer molecules were quite dissimilar.

It is in the above sense that the dyed fibres may behave as an ideal solution although the dye–fibre interaction almost certainly is specifically located. Merian has shown that a high degree of conformational agreement exists in certain cases between disperse dyes and polyester fibres which suggests the existence of dispersion force interactions. The work of Jones and Thompson suggests that in other cases, at least, hydrogen bond interactions may be involved (see Chapter 7).

III THREE-COMPONENT SYSTEMS

In the dyeing context simple three-component systems have been little studied. However the system in which dye and water vapour are adsorbed by hydrophobic substrates has been studied recently. Water vapour is adsorbed by most textile substrates in accordance with a BET isotherm although adsorption is accompanied in many cases by swelling. The latter precludes application of the BET isotherm equation in any sort of rigorous sense since v_m is not a constant. Jones and Thompson showed that water and dye molecules in the vapour phase competed for sites in cellulose acetate and that hysteresis effects were marked.

It is likely that the competition effects arose from one of two factors, conformational or specific bonding. Disperse dyes are normally weakly polar and possess hydrogen bonding capacity as does water. Competition for H-bonding sites is thus possible although the binding forces are somewhat large for such a hypothesis. The alternative is a conformational competition whereby water causes a conformational change in the substrate which disfavours dispersion force interaction between dye and polymer, and vice versa. The latter explanation is compatible with the observed hysteresis in desorption experiments where water vapour is allowed to displace dye and vice versa. Conformational competition is known in other dyeing situations as will be shown. It is characterized by a varying saturation value or maximum absorption value as the competing molecules vary in concentration. This is certainly to be seen in the dye–water vapour–cellulose acetate system in which the maximum adsorption value of the dye falls as the water content of the fibre is increased.

Solute–Solvent–Sorbate systems   These comprise the overwhelming majority of dye adsorption systems which have been studied. The use of dye solutions instead of dye vapours makes possible the use of ionizing dye molecules as well as the weakly polar disperse dyes, and a much wider variety of absorbing substrates requires to be considered, e.g. cotton, proteins, polyamides as well as polyester, cellulose acetate, polypropylene. Dyeing systems which have to be considered are often more complex. In practice the solvent which is used is water. The use of alternative solvents is a matter for much active research and separate consideration will be given to this aspect.

In dealing with the three-component solution system there will be few differences as compared with the vapour phase three-component dye system considered above. The system will be equivalent to one in which the water vapour pressure has reached saturation value and the concentration of dissolved dye will determine the dye uptake. Due to the competition effect the maximum adsorption value will be lower than in vapour phase dyeing. The linear partition isotherm would be expected to be followed. This is in general the situation which is observed experimentally. However there are complications due to physical changes in dye crystals during dyeing, aggregation phenomena and other effects in certain cases. Nevertheless the disperse dye–water–substrate system behaves in a generally ideal way. This is not true of systems using ionizing dyes and protein and other hydrophilic substrates.

The principles upon which discussion of the adsorption phenomenon can be based in these cases have been outlined but their manifestation in each case is sufficiently varied to warrant separate discussion in terms of specific systems.

IV   THE THERMODYNAMICS OF ADSORPTION EQUILIBRIA

The classical method of describing equilibria is in terms of thermodynamics. Accordingly at equilibrium in a system

$$\mu_s = \mu_s^0 + RT \ln a_s \qquad (2.68)$$

$$\mu_f = \mu_f^0 + RT \ln a_f \qquad (2.69)$$

in which μ_s and μ_f are the chemical potentials of the adsorbate in the external and adsorbant surfaces respectively; a_s and a_f are the corresponding activities; μ_s^0 and μ_f^0 are the corresponding standard state chemical potentials.

At equilibrium the chemical potentials in the two phases must be equal so that

$$\mu_s^0 + RT \ln a_s = \mu_f^0 + RT \ln a_f \qquad (2.70)$$

or

$$\mu_s^0 - \mu_f^0 = -\Delta\mu^0 = RT \ln a_f/a_s \qquad (2.71)$$

and the term $-\Delta\mu^0$ represents the change in standard chemical potential or standard molal free energy in the transfer from the external to the adsorbant phase and is thus the *affinity*.

This equation is of no experimental value however due to the abstract terms describing activity. In order to interpret the equation in terms of real forces etc. a model has to be constructed which tends to predetermine the physical interpretation to which the equation may lead. This means that several diverse theoretical approaches may all lead to the same final result. Thermodynamics consequently may enable accurate interpolations of data and useful indications of what may be important factors in the equilibrium situation but it can provide no information about molecular mechanisms.

In order to proceed with equation (2.71) it is necessary to come to some decision regarding the term, a_f, describing the activity of the adsorbate in the adsorbed state.

The free energy per mole, μ, is given by

$$\mu = H - T\Delta S = E + PV - TS \qquad (2.72)$$

Therefore

$$\left| \frac{\partial \mu}{\partial P} \right|_T = V \qquad (2.73)$$

Since $PV = RT$

$$\partial\mu = RT \, \partial P/P = RT \ln P \tag{2.74}$$

integration between μ and μ^o gives

$$\mu - \mu^0 = RT \, (\ln P - \ln P^0) \tag{2.75}$$

and defining the standard state as 1 atmosphere

$$\mu - \mu^0 = RT \ln P \tag{2.76}$$

For an ideal mixture of ideal gases the partial pressures are used and if n represents the number of moles of a component

$$P_i V = n_i RT \text{ for component } i \tag{2.77}$$

For each gas in the mixture

$$\mu_i - \mu_i^0 = RT \ln P_i \tag{2.78}$$

For n_i moles

$$n_i(\mu_i - \mu_i^0) = RT n_i \ln P_i \tag{2.79}$$

For a chemical reaction,

$$\Delta\mu^0 = \mu^0 \text{ (products)} - \mu^0 \text{ (reactants)}$$

because the μ terms are equal at equilibrium.
Therefore

$$-\Delta\mu^0 = RT \ln \sum n_i \ln (P_i)_e \tag{2.80}$$

where $_i$ represents the stoichometric number of moles, (negative for a reactant and positive for a product), and e signifies equilibrium values. Thus for

$$aA + bB \rightleftharpoons cC + dD$$

$$-\Delta\mu^0 = RT \ln (P_C)_e^c (P_D)_e^d / (P_A)_e^a (P_B)_e^b \tag{2.81}$$

$$-\Delta\mu^0 = RT \ln K \tag{2.82}$$

For one gas adsorption situation, K is the equilibrium constant for the adsorption and equal to

$$\frac{\Gamma}{[P(S - \Gamma)]} \tag{2.83}$$

in the Langmuir model.
Thus for specific site adsorption

$$-\Delta\mu^0 = RT \ln \frac{\Gamma}{[P(S - \Gamma)]} \tag{2.84}$$

So defining θ as $\Gamma.S$

$$-\Delta\mu^0 = RT \ln \frac{\theta}{(1-\theta)} \cdot \frac{1}{P} \tag{2.85}$$

Thus by analogy with the general expression

$$-\Delta\mu^0 = RT \ln\left(\frac{a_f}{a_s}\right) \tag{2.86}$$

the activity of the adsorbed species is $\theta/1-\theta$ and the equation inevitably relates to a Langmuir isotherm.

If on the other hand the partition isotherm had applied, i.e. $K = \Gamma/P$, then

$$-\Delta\mu^0 = RT \ln \Gamma/P. \tag{2.87}$$

The equation may be applied to the solute/solvent/substrate situation by employing concentrations instead of pressure and surface concentration.

The term $\theta/1-\theta$ represents the ratio of occupied to unoccupied sites and like a mol fraction is dimensionless. It is not a mol fraction but a probability or entropy term which arises from the assumption that adsorbed molecules do not migrate over the surface and can only be moved from one site to the next by a desorption/readsorption sequence. If that assumption is not made and molecules are mobile within the surface, the equation which is obtained is identical within equation 2.87 which fits the partition isotherm. The same condition also applies if $\theta < < 1$ or $-\Delta\mu^0[C_s]/RT < < 1$.

The free energy of adsorption as calculated using equation (2.85) or (2.87) or any other which might be devised on the basis of other models, is not a constant being a temperature dependent. Its numerical value is dependent upon the model used as has been shown. This is not true of the *enthalpy of adsorption*.

The three thermodynamic parameters are linked by

$$\Delta\mu^0 = \Delta H^0 - T\Delta S^0 \tag{2.88}$$

The entropy of adsorption is not susceptible to experimental determination being calculated from $\Delta\mu^0$ and ΔH^0 values. The numerical value of $\Delta\mu^0$ cannot be verified and the only parameter which can be independently determined is the enthalpy of adsorption. Combining equation (2.88) with (2.71) gives

$$-\Delta\mu^0 = RT \ln \frac{a_f}{a_s} = \Delta H^0 - T\Delta S^0 \tag{2.89}$$

Hence

$$\ln a_s = -\left(\frac{\Delta H^0}{R}\right) \cdot \left(\frac{1}{T}\right) + \frac{\Delta S}{R} + \ln a_f \tag{2.90}$$

If it is assumed that a_f varies only as C_f in accordance with $a_f = f(C_f)$ where $f(\ )$ implies a constant but unknown function independent of C_f then the selection of different temperatures giving a constant value of C_f should mean the comparison of situations in which a_f is a constant and

$$\ln a_s = -\left(\frac{\Delta H^0}{R}\right) \cdot \left(\frac{1}{T}\right) + k \qquad (2.91)$$

in which $k = \Delta S^0/R + \ln a_f$. We require to know either the values of the activity coefficient of the dye in the external phase at different temperatures, or that the activity coefficient does not vary with temperature so that

$$\ln [C_s] = -\left(\frac{\Delta H^0}{R}\right) \cdot \left(\frac{1}{T}\right) + k' \qquad (2.92)$$

in which $k' = \Delta S^0/R + \ln a_f/\gamma_s$.

The required information is easily abstracted from a series of experimental isotherms as shown in Fig. 2.3.

The assumptions of this treatment needs to be carefully considered in relation to any particular case since it is distinctly possible that $f(\ )$ is in certain cases temperature dependent due to special factors.

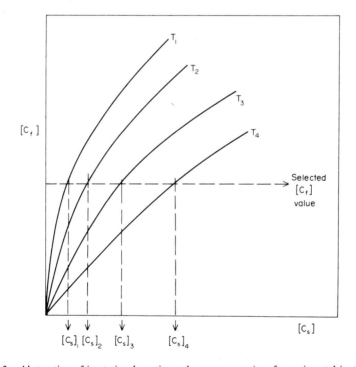

Fig. 2.3. Abstraction of isosteric adsorption values from a series of experimental isotherms.

From the free energy and enthalpy values the entropy of adsorption may be calculated but it should be remembered that the value obtained will incorporate the accumulated errors in $\Delta\mu^0$, T and ΔH^0 and will additionally vary according to the assumptions used in calculating the affinity.

V FURTHER READING

Adamson, A. W., *The Physical Chemistry of Surfaces*, Interscience (1960).

de Boer, J. H., *The Dynamical Character of Adsorption*, 2nd edition, Oxford University Press (1968).

Guggenheim, E. A. and Fowler, R. H., *Statistical Thermodynamics*, Cambridge University Press (1939).

Hayward, D. O. and Trapnell, B. M. W., *Chemisorption*, 3rd edition, Chapman and Hall (1966).

Moelwyn Hughes, E. A., *Physical Chemistry*, Pergamon Press (1957).

Shaw, D. J., *Introduction to Colloid and Surface Chemistry*, Butterworths (1968).

Thompson, J., Ph.D. Thesis, Leeds University (1969).

Chapter 3

Kinetics of Dyeing

I   INTRODUCTION

The process of dyeing consists essentially of two steps: the transport of dye from a bath (generally an aqueous solution) into a fabric or an array of fibres, and the subsequent uniform distribution of the dye molecules over the available dye-binding sites. These "macro" steps themselves can be subdivided into elementary processes:

i.   transport of dye from the bulk of the solution to the solution–fabric interface
ii. material transport across the interface
iii. migration of the dye molecules in the textile material
iv. interaction of the dye molecules with the binding sites in the substrate.

Depending upon the operating conditions, one or other of these elementary processes can assume a prevalent role in determining the outcome of the dyeing process and the quality of the final product. They are governed by different mechanisms and by manipulating the dyeing conditions (pH, temperature, salt concentration, liquor ratio, etc.) the relative contribution of an elementary process can be enhanced or diminished, and thus the course of the overall process regulated.

The efficient operation of a dye house requires a thorough understanding of the mechanisms involved. As many of the operational costs are proportional to time, one of the main objectives of the operator is to optimize the dyeing process with respect to time. The understanding of the factors which govern the kinetics of the dyeing process therefore assume special importance.

The kinetics of any process, and consequently that of dyeing, can be discussed by using:

i.   the phenomenological, or thermodynamic treatment
ii. the macroscopic kinetic discussion, or
iii. the molecular approach.

The phenomenological approach considers the general relationships that exist between the environmental parameters (e.g. temperature, pressure volume, salt concentration, liquor ratio, amounts of various materials) and the quantities that characterize the processes (rates or extents of reaction, etc.). However this treatment establishes general correlations and thus can solve problems on a general level only; it is unable to tackle specific, practical questions as it does not use data pertaining to the physical or chemical properties of materials.

The main purpose of the phenomenological treatment is to clarify the nature of the relationships between the various factors and to establish their relative importance. It's role is analogous to that of thermodynamics in the treatment of equilibrium processes.

The macroscopic–kinetic approach, on the other hand, approximates the real situation by using a simplified mathematical model, and then examines the processes in terms of a postulated model. The nature of the model depends upon the problem that it tries to approximate. Thus different models are suitable for describing the dyeing of single fibres, fibre arrays or fabrics.

This approach is suitable for solving practical problems as it uses actual, measurable quantities (e.g. diffusion coefficient) which are characteristic of the particular system under consideration and which can be determined experimentally.

Description of the dyeing process in quantitative terms however is not sufficient; often predictions on the performance of the dyes and their dyeing behaviour, on the basis of their molecular structures, are also necessary. Thus a molecular theory of dyeing is required which interprets measured macroscopic properties in terms of molecular quantities.

In the following, the three approaches to the kinetics of dyeing will be discussed in some detail.

II   THE PHENOMENOLOGICAL APPROACHES

A   An outline of the principles of the thermodynamics of irreversible processes

Until recently, only dyeing equilibria (free energy, enthalpy and entropy changes occurring during the interaction of dyes with substrates) were amenable to thermodynamic treatments. Equilibrium thermodynamics, however, does not include time among its variables, and therefore cannot treat rate processes.

During the last two decades a thermodynamic theory for rate processes has been developed and it is now possible to discuss time dependent processes by using the Thermodynamics of Irreversible Processes (T.I.P.). (For a detailed discussion of T.I.P. see Fitts, 1962; de Groot and Mazur, 1962; Katchalsky and Curran, 1965; Prigogine, 1967.)

Just as classical or equilibrium thermodynamics rests on fundamental assumptions, T.I.P. is based on a few postulates which confine it's use to rate processes near equilibrium states, which are relatively slow.

A convenient starting point for discussing the essential features of T.I.P. is the second law of thermodynamics. This law states that in a closed isothermic system, where no material exchange with the surrounding medium takes place, any spontaneously occurring process leads to an increase in the entropy of the system. Provided the change is infinitely slow, by always passing through equilibrium steps, the magnitude of the entropy change is given by

$$dS = \frac{dQ}{T} \tag{3.1}$$

where S is the entropy, Q the heat exchanged with the surrounding, and T the absolute temperature. (According to accepted convention, the heat absorbed is positive, the heat released is negative.)

If the same system undergoes change at finite rate, (it does not always pass through equilibria), then

$$dS > \frac{dQ}{T} \tag{3.2}$$

The difference between the entropy changes, denoted by (3.1) and (3.2), equals the entropy production that occurs in the system as a result of the irreversibility of the process. Consequently, the total entropy change accompanying an irreversible process can be expressed as the sum of two terms:

$$dS = \frac{dQ}{T} + dS_i$$

or
$$dS = dS_e + dS_i \tag{3.3}$$

where dS_e denotes the entropy change due to the exchange of heat with the surroundings during the process, and dS_i the entropy change due to the irreversibility of the process. The fundamental function which measures the irreversibility of a process or the rate of entropy production is ϕ, the dissipation function; it is defined as

$$\phi = \frac{T dS_i}{dt} \tag{3.4}$$

The theory of T.I.P. further postulates that the dissipation function can be expressed as a sum of products of fluxes (J_i) and conjugate forces (X_i), each of which represent an irreversible process in the system:

$$\phi = \sum_i J_i X_i \tag{3.5}$$

The forces and their conjugate fluxes can be of diverse nature. For instance, a temperature gradient as a force (X_i) induces a conjugate heat flux, an electric potential gradient as a force causes the flow of electric current as a flux, a chemical potential gradient as a force results in a diffusional flow of material as flux, a pressure gradient gives rise to a flow of material as a flux, an affinity difference brings about the flow of a chemical reaction, etc.

Considerable freedom is allowed in choosing the forces and the fluxes. However it is important that their products should have the dimensions of energy over time, the same as the dissipation function. (This latter rule can be used to test the self-consistency of the system.)

A third assumption of T.I.P., which confines its use to relatively slow processes, states that a linear relationship exists between the fluxes and all the forces which operate in a system. This postulate is based on the empirical observation that many of the phenomenological laws of physics are linear relationships. For instance, Ohm's law states a linear relationship between an electric current I and a potential gradient E;

$$I = R . E \tag{3.6}$$

Similarly, Fick's law of diffusion states an analogous relation between material flow and the concentration gradient:

$$J = -D\left(\frac{dC}{dx}\right) \tag{3.7}$$

where D is the diffusion coefficient, C the concentration of dye, x the length coordinate and J the material flux. In general, therefore, the relationship between the fluxes and the conjugate forces can be expressed by

$$J_i = \sum_j L_{ij} X_j \tag{3.8}$$

where L_{ij} terms are proportionality factors, and are known as phenomeno-logical coefficients.

If several irreversible processes take place simultaneously then each flux will depend on all the forces present and vice versa. Thus in a system where both a temperature and a concentration gradient exist at the same time, simultaneous heat and material flows will occur, each flow depending on both gradients (thermal diffusion). Similarly, in a system where electric potential gradient and heat gradient are both present, the electric current and the heat flow will depend on both gradients (thermal, electricity). The well known cross phenomena of physics arise in this way. For instance, in a fluid mixture a thermal gradient, in addition to a heat flow, will also give rise to material transport (thermal diffusion). Similarly, an electric potential gradient can cause, under certain circumstances, both an electric current and a heat flow (thermal electricity).

Thus the phenomenological coefficients can be described as straight coeffi-cients, correlating fluxes and conjugate forces (e.g. electric current with potential gradient, extent of chemical reaction with difference in chemical potential, diffusional flow with chemical potential gradient) or cross coeffi-cients, relating a particular flux with the non-conjugate forces (e.g. electric current with temperature gradient, diffusional flow with temperature gra-dients, etc).

A further postulate of T.I.P. states that the reciprocal cross phenomenologi-cal coefficients, those that relate the reciprocal fluxes and forces, are equal ($L_{ij} = L_{ji}$). Thus, for instance, the coefficient relating diffusional flow with temperature gradient and the coefficient relating heat flow with chemical potential gradient have the same value (Onsager's Reciprocal Relationship).

Finally we assume that, even though the system is undergoing an irrever-sible change in its entirety and consequently is not in thermodynamic equili-brium, limited domains of the system can be regarded as being in a quasi-equilibrium state (local equilibrium) and can therefore be characterized by means of classical equilibrium thermodynamic relationships. In particular, we assume that Gibbs' equation is valid within limited local domains of an irreversible system. Therefore for any point of the system the dissipation function can be expressed in terms of thermodynamic variables with the help of the Gibbs' equation

$$\phi = T\left(\frac{dS}{dt}\right) = \frac{d}{dt}(U + pV - \sum_i \mu_i\, n_i) \tag{3.9}$$

where U is the internal energy, p the pressure, V the volume, μ_i the chemical potential and n_i the number of moles of the ith type present in the system.

We can summarize the fundamental assumptions of the Thermodynamics of Irreversible Processes as follows:

i. In a closed system which undergoes irreversible processes the entropy always increases. The entropy is divisible into two terms, the entropy flow and the entropy production.

ii. The magnitude of the entropy production is measured by the dissipation function. The dissipation function can be expressed as a sum of the products of the fluxes representing the irreversible processes, and the conjugate forces causing the fluxes.

iii. A linear relationship exists between each flow and all the forces in the system.

iv. The matrix of the phenomenological coefficients is symmetrical, i.e. the reciprocal coefficients are equal $(L_{ij} = L_{ji})$.

v. A local thermodynamic equilibrium can exist at any point and time in the system.

B   Network thermodynamics

A serious limitation of T.I.P. is that it is not applicable to non-linear phenomena. To overcome this problem, Oster, Perelson and Katchalsky (1971) have recently suggested a new approach. Following some fundamental ideas, first put forward by Meixner (1966), these authors developed a phenomenological approach based on the theory of electrical networks.

Network thermodynamics is based essentially on the following arguments. Most irreversible processes can be expressed as a product of two variables: an effort variable e and a flow variable f. For instance, in diffusional processes—the chemical potential gradient and the mass flow; in chemical reactions—the affinity and the reaction rate, etc. The product of the two variables is an energy production rate (e times f) which is made up of two components: a utilizable energy (reversible energy or storage) and a dissipated energy (entropy production). The time integral of either of the two quantities is generally easier to measure by experiment than the variables themselves. In a diffusional process the time integral of the material flow is Δq, and the amount of material transported during time t

$$\Delta q(t) = \int_0^t e(t)f(t)\,\mathrm{d}t \qquad (3.10)$$

which is an easily measurable quantity.

Any system can be represented by a network of finite number modes and branches. The kinetic process which takes place in the system can then be characterized as a sum of products of e_i and f_i, representing the effort and flow variable respectively for the ith branch of the network.

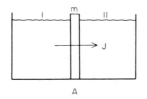

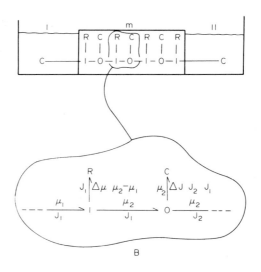

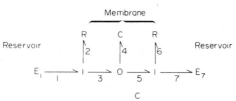

Fig. 3.1.   Schematic representation of transport of material through a membrane. For details see text. (Reproduced with permission from Oster et al, 1971.)

As an example, the transport of material from a Reservoir I into Reservoir II through a membrane can be represented by the following scheme (Fig. 3.1). The process through the membrane can be subdivided into three elements in series:

    i.   the flux across the reservoir I—membrane interface
   ii.   the flow inside the membrane, and,
  iii.   the flow across the second interface into reservoir II.

Each elementary process can be represented by two network elements in series: a dissipative process, i.e. a resistance

$$(\text{denoted by } \underset{-1-}{R})$$

and a storage process, or capacitance

$$(\text{denoted by } \underset{-0-}{C}).$$

The presence of a resistance and capacitance in series signify the fact that across the membrane, drops in both the chemical potential and the flow of material J occur. The values of the network elements, that of the resistance R_m and the capacitance C_m can be expressed in terms of measurable experimental quantities. Thus,

$$R_m = \frac{RT \, \Delta x}{D_m \, (N_i/V)} \tag{3.11}$$

and

$$C_m = \frac{(N_i/V)\Delta x}{RT} \tag{3.12}$$

where R_m is the membrane resistance, C_m is the membrane capacitance Δx is the thickness of the membrane, N_i is the number of moles of permeant in the membrane, and D_m is the diffusion coefficient of the permeant in the membrane.

The time course of the transport process across the membrane can then be described by an equation very similar to the one used to represent alternating electric currents, except that the electric potentials E are replaced by the chemical potentials:

$$\frac{(d\,\mu_m)}{dt} = \left[\frac{2}{(R_m \, C_m)}\right](\langle\mu\rangle - \mu_m)$$

$$= \frac{(\langle\mu\rangle - \mu_m)}{\tau_m} \tag{3.13}$$

where $\langle\mu\rangle$, the mean chemical potential in the system is defined as

$$\langle\mu\rangle = \frac{(\mu^I + \mu^{II})}{2} \tag{3.14}$$

and τ_m, the relaxation time of the process by

$$\tau_m = \frac{R_m \, C_m}{2} \tag{3.15}$$

III   APPLICATION OF T.I.P. TO DYEING

A   Dyeing with uncharged compounds

Dyeing of materials involves the transport of dyes from a reservoir (e.g. the dye bath) into a substrate (e.g. a textile material). The process is relatively slow near to the equilibrium and therefore can be treated by T.I.P. In recent years a comprehensive discussion of dyeing in terms of T.I.P. has been published (Breuer, 1965; McGregor, 1966; Milicevic, 1969, 1970; Milicevic and McGregor, 1966a). Milicevic and McGregor (1966b) used the following approach.

The dye bath and the substrate respectively are regarded as two homogenous compartments: bath (b) and fibre (f), separated by an infinitely thin membrane (Fig. 3.2).

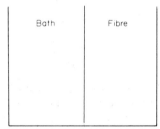

Bath   Fibre

Fig. 3.2.   A schematic representation of a two compartment model for the dyeing system.

A thermal equilibrium is assumed.

$$T_b = T_f = T \tag{3.16}$$

The system, containing the three components, dye (1), water (2) and textile material (3), can then be characterized by

$$J_1 = \left(\frac{1}{T}\right)(L_{11}A_1 + L_{12}A_2 + L_{13}A_3)$$

$$J_2 = \left(\frac{1}{T}\right)(L_{21}A_1 + L_{22}A_2 + L_{23}A_3)$$

$$J_3 = \left(\frac{1}{T}\right)(L_{31}A_1 + L_{32}A_2 + L_{33}A_3) \tag{3.17}$$

where

$$A_i = \Delta\mu_i + \bar{V}_i\Delta p$$

and

$$\Delta\mu_i = \mu_{i,b} - \mu_{i,f} \tag{3.18}$$

in which μ is the chemical potential, $\bar{V}$ is the partial molal volume and p the pressure.

If we assume that no flow of substrate takes place across the phase boundary, then

$$J_3 = O \tag{3.19}$$

The pressure difference can then be expressed as

$$\Delta p = -\frac{\sum_i L_{3i} \, \Delta\mu_i}{\sum_i L_{3i} \, \bar{V}_i} \tag{3.20}$$

Furthermore we can assume that before the dyeing process starts, as the diffusion of the solvent is very much faster than that of the dye, an equilibrium with respect to the solvent has already been established.

$$A_2 = O \tag{3.21}$$

therefore (3.17) can be simplified

$$J_1 = \left(L_{11} - L_{13} \frac{L_{31}}{L_{33}}\right) \frac{A_1}{T}$$

$$J_2 = \left(L_{21} - L_{23} \frac{L_{31}}{L_{33}}\right) \frac{A_1}{T} \tag{3.22}$$

If the total volume changes as a consequence of the dyeing is zero, then the volume flow becomes zero:

$$J_v = \sum_i \bar{V}_i J_i = 0 \tag{3.23}$$

and consequently

$$J_2 = -\left(\frac{\bar{V}_1}{\bar{V}_2}\right) J_1 \tag{3.24}$$

Denoting

$$L = \left[L_{11} - L_{13}\left(\frac{L_{31}}{L_{33}}\right)\right] \tag{3.25}$$

the dye flow is then given by

$$J_1 = \left(\frac{L}{T}\right)(\bar{V}_1 \Delta p + \Delta\mu_1) \tag{3.26}$$

and the solute flow by

$$J_2 = -\left(\frac{L}{T}\right)\left(\frac{\bar{V}_1}{\bar{V}_2}\right)(\bar{V}_1 \Delta p + \Delta\mu_1) \tag{3.27}$$

Furthermore the term $\bar{V}_1 \Delta p$ is small for most cases and can be neglected, then

$$J_1 = \left(\frac{L}{T}\right)\Delta\mu_1 \tag{3.28}$$

We can extend the above treatment to include more than three components either by assuming that the material flows depend upon all the concentration gradients or, alternatively, neglecting the interactions of flows by assuming that $L_{ij} = 0$.

B Dyeing with charged compounds

The majority of dyes used in the dyeing of textiles and other materials (paper, leather, hair) are either acids or bases (electrolytes). If electrically charged compounds are transferred from one phase into another, the rate of transport will depend upon the electric potential that exists between the two phases. Consequently, in thermodynamic treatments of transport processes the chemical potential μ_i has to be replaced by the electro-chemical potential $\tilde{\mu}_i$. For a cation, we can express the electrochemical potential as

$$\tilde{\mu}_i^+ = \mu_i^+ + Z_i\,\Delta\psi F \tag{3.29}$$

where μ_i^+ is the chemical potential, $\tilde{\mu}_i^+$ the electrochemical potential of the cation or anion, Z_i the electrochemical equivalent, $\Delta\psi$ the electric potential difference between the two phases and F the Faraday constant.

The phenomenological equations can be written for each charged species separately

$$J_i = \sum_i L_{ij}\,\Delta\tilde{\mu}_j \tag{3.30}$$

However, electroneutrality requires that no net electric current, I should be transferred as a consequence of the dyeing process,

$$I = F\sum Z_i J_i = 0 \tag{3.31}$$

For a system composed of three different ionic species and an uncharged substrate, i.e. dye-ion (component 1), solvent (component 2) a salt ion (component 3) and a common counterion (component 4), equations (3.29) to (3.31) can be solved to give the values of electrical potential difference between the two phases in terms of the phenomenological coefficients of the ionic species:

$$-\Delta\psi = \frac{(L_{31}\,\bar{V}_1 + L_{33}\,\bar{V}_3 + L_{34}\,\bar{V}_4)\Delta p + L_{31}\,\Delta\mu_1 + L_{33}\,\Delta\mu_3 + L_{34}\,\Delta\mu_4}{F(L_{31}\,Z_1 + L_{33}\,Z_3 + L_{34}\,Z_4)} \tag{3.32}$$

The value of L, the overall phenomenological coefficient of the neutral dye-salt is given as

$$L = \frac{(L_{11} L_{33} - L_{13})^2 (L_{44} L_{33} - L_{34}^2) - (L_{14} L_{33} - L_{13} L_{34}^2)}{T L_{33} [(L_{44} L_{33} - L_{34}^2) + (L_{14} L_{33} - L_{13} L_{34})(Z_1/Z_4)]} \quad (3.33)$$

provided the value of L as given by (3.33) is used, (3.28) is also valid for neutral dye salts.

C   Integration of the flux equation (equation 3.28)

We can obtain useful information concerning the environmental factors which determine dye uptake by integrating (3.28). Before integration however, (3.28) has to be put in a suitable form by expressing the chemical potentials as a function of the dye concentrations; a model is required which adequately represents the physico-chemical properties of the dye solutions in the two compartments b (dye bath) and f (fabric).

We may regard both the dye bath and the dyed material as ideal solutions and use the well known functional relationship

$$\mu = \mu_0 + RT \log C \quad (3.34)$$

where C is the concentration of dye.

Alternatively, we may assume that the dye bath is an ideal dye solution and that in the dyed material the dye molecules are adsorbed on discrete binding sites. In the latter case functional dependence of the chemical potential on the dye concentration in the fabric depends on the nature of the binding isotherm. Thus, if the dye adsorption isotherm is linear, the chemical potential of the dye in the fabric is given by

$$\mu_f = \lambda C_f \quad (3.35)$$

where λ is the partition coefficient of the dye between the fabric and the bath and C_f the dye concentration of the fabric. For a Langmuir-type of dye binding equilibrium, when the dye binding occurs on discrete sites

$$\mu_f = \frac{C_f}{[K (C_{f, max} - C_f)]} \quad (3.36)$$

where $C_{f, max}$ is the maximum dye uptake possible.

For the safe of simplicity only the hypothetical case will be discussed. When the chemical potential of the dye in both the solution and the fabric can be expressed by (3.34), the flux is then obtained by using (3.28),

$$J_1 = \left(\frac{L}{T}\right)(\mu_b - \mu_f)$$

$$= LR \left[\mu_{b,0} - \mu_{f,0} + \ln\left(\frac{C_b}{C_f}\right) \right] \quad (3.37)$$

Assume that before the beginning of the dyeing experiment, when $t = 0$ $C_b = 0$, no dye is present in the fabric (compartment f):

$$C_b = \frac{(n_0 - n)}{V_b} = \frac{[(n_0 - n)\rho_b]}{m_b} \tag{3.38}$$

$$C_f = \frac{n}{V_f} \cdot \frac{n\rho_f}{m_f} \tag{3.39}$$

where n is the amount of dye transferred to compartment b (at $t = 0, n = 0$), n_0 the total amount of dye present in the system, V_b the volume of bath, V_f the volume of fibre, m_b the mass of bath, m_f the mass of fibre, ρ_b the density of bath, and ρ_f the density of fibre.

Express the flux of dye into f as

$$J = \left(\frac{dn}{dt}\right)\left(\frac{1}{Q}\right) \tag{3.40}$$

where Q is the surface area of the interfaces between the two compartments. The differences in the standard state are defined by

$$\mu_{b,0} - \mu_{f,0} = RT \ln K \tag{3.41}$$

where K is the intrinsic partition of the dye between the bath and the fabric, so

$$K = \frac{C_{f,\infty}}{C_{b,\infty}}$$

The liquor ratio, r is

$$r = \frac{V_b}{V_f} = \left[\frac{m_b \, \rho_f}{m_f \, \rho_b}\right] \tag{3.42}$$

and expressing the concentration in relative units

$$C = \frac{C_f}{C_{f,\infty}} \tag{3.43}$$

then combination of (3.37) to (3.42) yields

$$-\frac{dC}{dt} = B \ln\left[\frac{C}{K/r(1-C)+1}\right] \tag{3.44}$$

where

$$B = \frac{LRQ}{n_\infty} \tag{3.45}$$

Equation (3.44) gives the important relation between the rate of dyeing as a function of the total amount of dye used, the affinity of the dye to the fabric and the liquor ratio. To obtain the exact functional relationships, (3.44) has to be solved. However, it can only be integrated by numerical methods. In

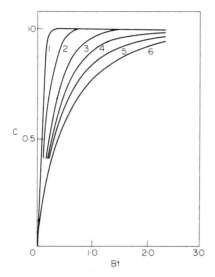

Fig. 3.3. The dye uptake, C, as a function of time calculated by equation 3.44. Curves 1, E = 1·99%; 2, E = 2·90%; 3, E = 3·70%; 4, E = 4·50%; 5, E = 5·30%. (Reproduced with permission from Milicevic and McGregor, 1966.)

Fig. 3.3 the value of C, the relative concentration of the dye in the fibre, is expressed as a function of the dimensionless quantity Bt (i.e. time scaled relative to the rate constant B) for various degrees of E, the final exhaustion. E is defined by

$$E = \frac{100}{1 + r/K} \%$$ (3.46)

The larger the value of E, the steeper the curve and the faster the build up of the dye concentration in the fibre.

D Some practical applications of rate equations (3.40–3.45)

Beyond its theoretical importance, (3.44) can be used for solving a number of practical problems which the dyer might encounter in day to day practice. For instance, it is important to calculate the time required to obtain a given shade (or dye uptake). If we define the degree of dye saturation as

$$\theta = \frac{n}{n_\infty}$$ (3.47)

equation (3.44) can be rearranged and integrated numerically, and the quantity θ expressed as a function of the three experimentally variables t, r and n_0

$$\theta = f(t, r, n_0) \tag{3.48}$$

Provided that the affinity K is determined in a separate experiment, the dyer will have considerable freedom to choose the best conditions (e.g. time of dyeing, temperature, etc.) compatible with his requirements.

For instance, consider the case when a particular shade requires a mixture of three dyes, with dye saturation values of θ_1, θ_2 and θ_3. The final saturation value for each of the dyes will be a function of time (t), the liquor ratio (r) and of the amount of dye initially entered into the dye bath ($n_{0,i}$)

$$\theta_1 = f_1(r, t, n_{0,1})$$

$$\theta^2 = f_2(r, t, n_{0,2})$$

$$\theta_3 = f_3(r, t, n_{0,3}) \tag{3.49}$$

Equation (3.44) shows that the dyer will have only a limited amount of freedom in choosing his initial dye composition, and once this is chosen either the liquor ratio or the dyeing time will be, *a priori*, determined.

Equation (3.44) can also be used to improve the economics of the operation. To illustrate this point, consider the case when the cost of dyeing process has to be minimized. The COST is a function of the amount of dye used and the operating costs. Assume the latter to be dependent on the time of dyeing and the liquor ratio (these two quantities will directly determine the various operating and plant cost),

$$\mathrm{COST} = F(n_{0,1}, n_{0,2}, n_{0,3}, r, t) \tag{3.50}$$

Out of the five variables in (3.50) only two are independent if a shade is chosen, for the values of θ_1, θ_2 and θ_3 are automatically fixed.

E   Dyeing with coupled fluxes

The addition of sparingly soluble organic solvents, e.g. benzyl alcohol, butanol, to a dye bath enhances in certain cases the dye uptake (Gorkhale *et al.*, 1958; Karrholm and Lindberg, 1956; Peters and Stevens, 1958). The effect of butanol on the rate of uptake of various acid dyes by wool is illustrated in Fig. 3.4. In certain cases it appears that the role of the solvent is only to remove

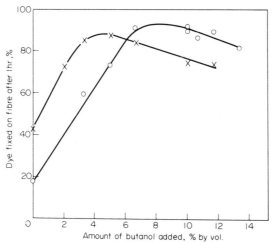

Fig. 3.4. Effect of different concentrations of n-Butanol on dye exhaustion after 1 hr.
(Reproduced with permission from Gorkhale *et al.*, 1958.)

a layer which acts as a diffusion barrier from the surface of the fibre (Butcher
and Cussler, 1972). In general however, the increased dye uptake is due to an
accelerated diffusion process which occurs as a result of a coupling of fluxes
between the dye and the organic solvent (Cussler and Breuer, 1972a, 1972b).

The diffusion of a dye in a mixed solvent is described by the use of ternary
flux equation (Cussler and Breuer, 1972a, 1972b)

$$- J_d = D_d \nabla C_d + D_{d,s} \nabla C_s \qquad (3.51)$$

where J_d is the flux of the dye, ∇C_d the concentration gradient of the dye, ∇C_s
the concentration gradient of solvent, D_d the "straight" diffusion coefficient
of the dye, and $D_{d,s}$ the "cross" diffusion coefficient of the dye.

The magnitude of both D_d and $D_{d,s}$ will be strongly dependent on the
thermodynamic activity of all the components present in the system, and
their values can be expressed as:

$$D_d = \frac{DC_d}{RT} \left(\frac{d\mu_d}{dC_d} \right)_{C_s} \qquad (3.52)$$

and

$$D_{d,s} = \frac{DC_d}{RT} \left(\frac{d\mu_s}{dC_d} \right)_{C_s} \qquad (3.53)$$

where D is the binary diffusion coefficient, μ_d and μ_s the chemical potentials
of the dye and of the solvent.

Using various simplifying assumptions, (3.51) can be reduced to a simpler form (Cussler and Breuer, 1972a, 1972b)

$$-J_d = D\nabla C_d + \frac{DC_d}{C_s^0}\ln\left[\frac{C_d^s(S)}{C_d^s(W)}\right]\nabla C_s \tag{3.54}$$

$$= D\nabla C_d + D'\nabla C_s \tag{3.55}$$

where C_s^0 is the molarity of the pure solvent, $C_d^s(S)$ the concentration of the dye at the solubility limit in the organic solvent, and $C_d^s(W)$ the concentration of the dye in pure water at its solubility.

Using NaCl as a model, Cussler and Breuer (1972a; 1972b) determined the value of D' in various aqueous organic solvent mixtures and plotted it as a function of its solubility in the solvent. A nearly nine fold increase in the value of the effective diffusion coefficient D' was observed in certain aqueous organic solvent mixtures (Fig. 3.5).

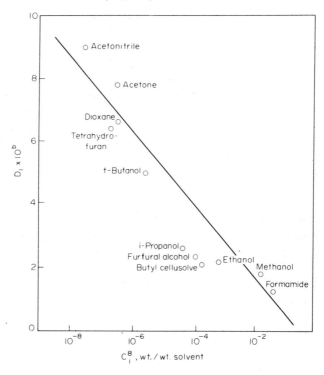

Fig. 3.5. Pseudo-binary diffusion coefficient (D_1) as a function of solubility (C_1^s) in non-solvent. (Reproduced with permission from Cussler and Breuer, 1972a.)

The coupling of diffusional flows also influences the rate of dye uptake in simple aqueous systems, i.e. in the absence of any organic solvent. Depending on the condition applied, parallel or opposing flows of dye and water can occur during dyeing of fibres. The relative directions and magnitudes of the various material flows will, in the end, determine the actual rate of dye uptake. A theory (Breuer, 1965) for calculating these effects was found to give good agreement with experimental results. As shown in Fig. 3.6 the

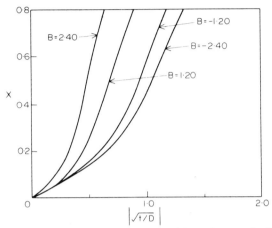

Fig. 3.6. Plot of the relative dye uptake as a function of time. The quantity B depends on the relative directions and magnitudes of dye and solvent flows. (Reproduced with permission from Breuer, 1965.)

value of X, the relative dye uptake at any given time (i.e. relative to the maximum equilibrium dye uptake), depends on the value of a parameter denoted B. The quantity B is positive if the directions of dye and water flows coincide and is negative when they are in an opposite direction. The differences between the rate of dyeing owing to flux coupling can be fairly substantial.

IV   THE DIFFUSION EQUATION

The thermodynamic treatment of transport phenomena cannot predict exact transport rates on the basis of the chemical structure of the dyes and of the substrate. Predictive calculations of this type generally require the knowledge of the diffusion coefficient of the dye-substrate system and the geometry of the substrate. Historically, the diffusion coefficient originates from the basic mathematical law of diffusion, first formulated by Fick who postulated that heat and material transport processes are analogous and are governed by similar physical laws. (For further reading see Crank, 1956; Jost, 1952; Tyrrell, 1961). According to Fick, J, the flux material (amount of material transported through a unit cross sectional area normal to the direction of flow), is opposite in its direction (hence the negative sign) and proportional to the concentration gradient:

$$J = -D \operatorname{grad} C \qquad (3.56)$$

For the particular case of one dimensional material transport, using cartesian coordinates

$$J = -D\left(\frac{dC}{dx}\right) \qquad (3.57)$$

This is known as Fick's law.

The quantity D, the factor of proportionality, is called the diffusion coefficient. Since the dimensions of flux, J, are $\text{kg}^{-2}\,\text{sec}^{-1}\,\text{m}^{-2}$ and that of concentration gradient $\text{kg}\,\text{m}^{-4}$, the diffusion coefficient has dimensions of $\text{m}^2\,\text{sec}^{-1}$.

A   The differential equation of diffusion

Fick's first law postulates a linear relationship between the material flux and the concentration gradient. However, it does not give the solution to the most important questions, notably the value of the existing concentration at a given point in space and time. This question can only be answered when the general differential equation of diffusion has been formulated and solved.

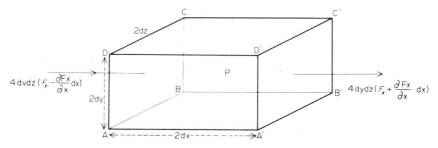

Fig. 3.7.   Schematic presentation of a volume element in the diffusion medium. (Reproduced with permission from Crank, 1956.)

Consider a volume element (Fig. 3.7) a parallelepiped through which material flow occurs. The flux and the concentration at the centre of the volume elements, at point x are denoted by J_x and C_x respectively. The rate by which the substance enters into the volume element through plane at the plane $x - dx$ can be calculated by expanding J_x into a Taylor-series around its values at x,

$$-4\,dy\,dz\left[J_x - \left(\frac{\partial J_x}{\partial x}\right)dx\right]dt \qquad (3.58)$$

Similarly, the loss through the plane at $x + dx$ is given by

$$-4\,dy\,dz\left[J_x + \left(\frac{\partial J_x}{\partial x}\right)dx\right]dt \qquad (3.59)$$

The difference between the two fluxes is the net gain of material in the volume element during time dt

$$-8\,\mathrm{d}y\,\mathrm{d}x\,\mathrm{d}z\left(\frac{\partial J_z}{\partial z}\right)\mathrm{d}t \tag{3.60}$$

The fluxes through the other faces of the parallelepiped can be calculated in a similar way giving the flow along the z axis as

$$-8\,\mathrm{d}y\,\mathrm{d}x\,\mathrm{d}z\left(\frac{\partial J_x}{\partial x}\right)\mathrm{d}t \tag{3.61}$$

and the flow along the y axis as

$$-8\,\mathrm{d}y\,\mathrm{d}x\,\mathrm{d}z\left(\frac{\partial J_y}{\partial y}\right)\mathrm{d}t \tag{3.62}$$

On the other hand the net increase in the amount of material in the parallelepiped during dt time period is also given by

$$8\,\mathrm{d}x\,\mathrm{d}y\,\mathrm{d}z\,\delta C_x \tag{3.63}$$

where ∂C_x is the increase in the concentration of the parallelepiped bounded by $2\,\mathrm{d}x$, $2\,\mathrm{d}y$ and $2\,\mathrm{d}z$.

Equating (3.63) with the sum of (3.60) to (3.62) gives

$$-\left(\frac{\partial C}{\partial t}\right)_x = \left(\frac{\partial J_x}{\partial x}\right)+\left(\frac{\partial J_y}{\partial y}\right)+\left(\frac{\partial J_z}{\partial z}\right) \tag{3.64}$$

but according to Fick's law

$$J_x = -D\left(\frac{\partial C_x}{\partial x}\right) \tag{3.65}$$

Combining (3.64) and (3.65)

$$\frac{\partial C}{\partial t}=D\left(\frac{\partial^2 C}{\partial x^2}+\frac{\partial^2 C}{\partial y^2}+\frac{\partial^2 C}{\partial z^2}\right) \tag{3.66}$$

or in a more general case where D is also a function of concentration

$$\frac{\partial C}{\partial t}=\frac{\partial}{\partial x}\left(D\frac{\partial C}{\partial x}\right)+\frac{\partial}{\partial y}\left(D\frac{\partial C}{\partial y}\right)+\frac{\partial}{\partial z}\left(D\frac{\partial C}{\partial z}\right) \tag{3.67}$$

The above equation is written for a Carthesian coordinate system. In dyeing problems, where generally fibres with a cylindrical symmetry are treated, it is often advantageous to use cylindrical coordinates (Fig. 3.8).

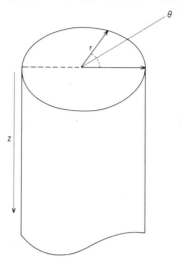

Fig. 3.8.   Schematic representation of the cylindrical coordinate system.

The diffusion equation then becomes:

$$\frac{\partial C}{\partial t} = \frac{D}{r}\left[\frac{\partial}{\partial r}\left(r\frac{\partial C}{\partial r}\right) + \frac{\partial^2 C}{r\partial\theta^2} + \frac{r\partial^2 C}{\partial z^2}\right]$$

(3.68)

where r, is the radius vector, θ its direction angle and z the length coordinate.

When the geometry of the system has not been defined it is often advantageous to use the vector notation,

$$\frac{\partial C}{\partial t} = \text{div}\,(D\,\text{grad}\,C) = \nabla(D\,\nabla c)$$

(3.69)

The diffusion is a partial differential equation, where the dependent variable C is a function of two independent variables x and t. The solution of the differential equation will give C as a function of t and x. The value of the constant of integration can only be obtained if the boundary conditions, which depend on the geometry of the system and the physical conditions under which the dyeing is occurring, are first defined.

B   Solution of the differential equation of diffusion

A standard method for solving partial differential equations is to assume that the independent variables are separable (Crank, 1956; Jost, 1952). Thus if we attempt to find a solution for the diffusion equation we can assume that the concentration is a function of a product of two functions, each of which is a function of only one independent variable:

$$C = X(x)\,T(t)$$

(3.70)

where $X(x)$ and $T(t)$ are assumed to be functions only of x and t respectively. Consequently, in solving (3.66) we obtain:

$$\left(\frac{\partial C}{\partial t}\right)_x = X \frac{dT}{dt} \tag{3.71}$$

$$\left(\frac{\partial C}{\partial y}\right)_t = T \frac{dX}{dx} \tag{3.72}$$

substitute (3.71) and (3.72) into equation (3.66), then

$$X \frac{dT}{dt} = DT \frac{d^2 X}{dx^2} \tag{3.73}$$

or

$$\frac{1}{T} \frac{dT}{dt} = \frac{D}{X} \frac{d^2 X}{dx^2} = -\lambda^2 D \tag{3.74}$$

where λ^2 is a constant.

Both sides of (3.74) can now be integrated giving respectively

$$T = \exp\left(-\lambda^2 Dt\right) \tag{3.75}$$

and

$$X = A \sin \lambda x + B \cos \lambda x \tag{3.76}$$

Equations (3.75) and (3.76) can be verified by differentiation.

Since the diffusion equation is a linear equation, the general solution is the combination of particular solutions. Consequently the general solution for (3.66) is given by:

$$C = \sum_{m=1}^{\infty} \left(\lambda_m \sin \lambda_m x_m + B_m \cos \lambda_m x_m\right) e - \lambda_m^2 Dt \tag{3.77}$$

where A_m, B_m and λ_m are constants which depend on the boundary conditions.

An alternative approach for solving the diffusion equation is by Laplace transforms. This essentially involves the conversion of a partial differential equation to a total differential equation, (i.e. the dependant variable only depends on one independent variable) and its subsequent solution by well known techniques. (See Crank, 1956; Jaeger, 1949; Jost, 1952.)

C Solution of the diffusion equation for a concentration dependent D

When D is a function of C the diffusion equation becomes non-linear and cannot be solved analytically. To overcome this difficulty Boltzmann (1894) postulated that the two independent variables in the diffusion equation, x and t, have a constant relationship and can be replaced, by a new variable y where

$$y = \frac{x}{2t^{1/2}} \qquad (3.78)$$

The diffusion equation will become

$$2y \frac{dc}{dy} = \frac{d}{dy}\left(D \frac{dc}{dy}\right) \qquad (3.79)$$

or, after integrating,

$$D(c) = 2 \frac{dy}{dc} \int_0^c 2y \, dc \qquad (3.80)$$

Resubstituting (3.78) into (3.79) yields

$$D(c) = \frac{1}{2t}\left(\frac{dx}{dc}\right) \int_0^c x \, dc \qquad (3.81)$$

The integral can generally be evaluated by numerical methods.

D Some analytical solutions of the diffusion equation

Having obtained experimental measurements correlating either the distribution of dye concentration or, alternatively the total dye uptake with the time of dyeing, the next step is to evaluate the diffusion coefficient from the experimental data. To do this, an analytical, exact solution of the diffusion equation is required.

A number of the most frequently used solutions of the diffusion equation will be reviewed. In all these cases the diffusion coefficient is assumed to be independent of C, t and x.

Generally two types of experimental data are available: either the dye concentration C is obtained as a function of a space coordinate, or the amount of dye uptake, M is measured as a function of time. To treat these two types of data, we require two different solutions for a given geometric arrangement.

Generally the substrate that is dyed consists of fibres and can be regarded in a first approximation as an assembly of infinite, isotropic cylinders. Therefore, in most cases of interest to the dyer the diffusion equation expressed in terms of a cylindrical coordinate system is used. The direction of dye penetration occurs from the periphery towards the centre of the cylinder, in a radial direction. In general, two particular solutions are of interest to the dyers: firstly when the dye bath can be regarded as being infinite, i.e. the dye concentration in the bath stays constant at its initial value during the entire time of dyeing irrespective of the amount of dye absorbed by the substrate; and secondly, the case of changing dye concentration in the bath (finite dye bath volume), when the amount of dye in the entire dyeing system (i.e. bath plus substrate) is fixed, i.e. at any time the dye extracted from the bath equals amount of dye taken up by the substrate.

The diffusion equation in cylindrical coordinates is

$$\frac{\partial C}{\partial t} = \frac{1}{r}\left[\frac{\partial}{\partial r}\left(rD\frac{\partial C}{\partial r}\right) + \frac{\partial}{\partial \theta}\left(\frac{D}{r}\frac{\partial C}{\partial \theta}\right) + \frac{\partial}{\partial z}\left(rD\frac{\partial C}{\partial z}\right)\right] \tag{3.82}$$

The last term, in accordance with the assumption of an infinitely long fibre, is zero. Equation (3.82) can be solved by the Laplace transform method.

In an infinite dye bath, the concentration of the dye at the surface stays constant during the entire dyeing experiment ($C_0 = $ const). The concentration distribution in the fibre is then

$$\frac{C - C_1}{C_0 - C_1} = 1 - \frac{2}{a}\sum_{n=1}^{\infty}\frac{\exp(-D\alpha_n^2 t)J_0(r\alpha_n)}{\alpha_n J_1(a\alpha_n)} \tag{3.83}$$

where C is the concentration at a given point along the radius, C_0 the initial dye concentration in the bath, C_1 the initial dye concentration in the fibre, J_0 and J_1 are Bessel Functions of zero and the first order, α_n their positive roots, and a the radius of the fibre. The appropriate concentration distributions in the fibre are shown in Fig. 3.9. The equation for the total dye uptake is

$$\frac{M}{M_\infty} = \frac{4}{\pi^{1/2}}\left[\left(\frac{Dt}{a^2}\right)^{1/2}\right] - \frac{Dt}{a^2} + \cdots \tag{3.84}$$

where M is the equilibrium dye uptake. The curve in Fig. 3.10 shows the computed uptake curves for an infinite bath.

In the case of the finite bath the amount of the dye present in the bath and in the fibre stays constant. At infinite time the dye distribution between the fibres and bath is governed by the equilibrium partition coefficient

$$\beta = \frac{M_{b,\infty}}{M_{f,\infty}} \tag{3.85}$$

where $M_{b,\infty}$ is the amount in the bath at $t = \infty$ and $M_{f,\infty}$ the amount in the fibre at $t = \infty$.

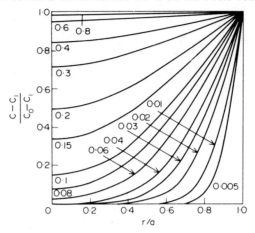

Fig. 3.9.   Concentration distribution of dyes in a cylindrical fibre at various times. C_1 = initial concentration in the fibre; C_0 = surface concentration; C = Concentration at a given time; a = radius of fibre. Numbers on curves are values of Dt/a^2. (Reproduced with permission from Crank, 1956.)

The dye uptake for small values of β where $1 \gg \beta$, small bath fibre ratio and/or high dye affinity is

$$\frac{M}{M_\infty} = \frac{1+\beta}{1+\frac{1}{4}\beta}\left[1 - \frac{\beta\,(Dt/a^2)^{-1/2}}{2\pi^{1/2}(1+\frac{1}{4}\beta)} + \cdots\right] \tag{3.86}$$

and for large value of β, i.e. $1 \ll \beta$, large bath fibre/ratio and/or small dye affinity

$$\frac{M}{M_\infty} = \frac{2(1+\beta)}{\beta}\left[\frac{2}{\pi^{1/2}}\left(\frac{Dt}{a^2}\right)^{1/2} - \left(\frac{1}{2}+\frac{2}{\beta}\right)\frac{Dt}{a^2} + \cdots\right] \tag{3.87}$$

In most cases of practical significance, however, the diffusion coefficient can be calculated from dye uptake by retaining respectively only the first term of (3.86) or (3.87)

$$\frac{M}{M_\infty} \propto \left(\frac{Dt}{a^2}\right)^{-1/2} \quad \text{or} \quad \frac{M}{M_\infty} \propto \left(\frac{Dt}{a^2}\right)^{1/2} \tag{3.88}$$

V   THE EXPERIMENTAL DETERMINATION OF THE DIFFUSION COEFFICIENT OF DYES

The evaluation of the diffusion coefficient from experimental data requires the solution of the diffusion equation for well defined boundary conditions. Generally two types of data are obtained in a dyeing experiment: either the concentration of the dye is measured as a function of space coordinates (e.g. thickness of a slab, or radius of a fibre) or, alternatively, the amount of dye uptake by a fibre fabric or polymeric membrane is obtained as a function of time. In either case, the computation of the diffusion coefficient requires the exact knowledge of the geometry of the sample. Some of the available experimental methods for the determination of the diffusion coefficient will be reviewed.

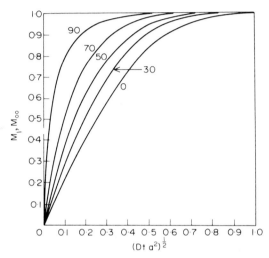

Fig. 3.10.   Dye uptake curves for a cylindrical fibre from a limited dye-bath. Numbers on the curves show the percentage of total solute finally taken up by the cylinder. (Reproduced with permission from Crank, 1956.)

A   Stationary diffusion method

This was first used by Garvie and Neale (1938) who studied the diffusion of Sky Blue FF through a sheet of cellophane using the following apparatus (Fig. 3.11).

Two compartments A and B containing well stirred solutions are separated by a thin film of a polymeric material. Initially only compartment A contains a dye. In compartment B the dye concentration is monitored as a function of time, and the amount of dye diffusing through the film is determined in this way. Vigorous stirring is maintained all the time. As the volumes of both compartments are very large we can assume that the dye concentration in A remains constant. The diffusion coefficient can then be calculated in two ways. Firstly the dye concentration in B is plotted against time; when the stationary state has been reached, the slope of the curve will be constant (Fig. 3.12).

$$\frac{\partial C}{\partial t} = \text{const} \qquad (3.89)$$

therefore

$$\frac{\partial}{\partial x}\left(D\frac{\partial C}{\partial x}\right) = 0 \tag{3.90}$$

and

$$D\frac{\partial C}{\partial x} = \text{const} = \frac{\Delta Q_t}{\Delta t\, F} \tag{3.91}$$

where F is the surface area of the membrane. The value for D can be expressed as integrating the left hand side of the equation

$$D = \text{const}\ \frac{x_A - x_B}{C_A - C_B} = \frac{\Delta x}{C_A - C_B}\,\text{const} \tag{3.92}$$

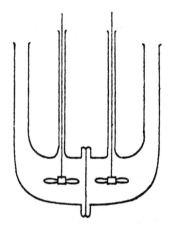

Fig. 3.11.   Schematic Representation of apparatus used for studying dye diffusion through polymeric fibres. (Reproduced with permission from Garvie and Neale, 1938.)

In a non-stationary state, when $t \neq 0$, $c_B \neq 0$, the amount of dye Q diffusing through the membrane is (Jost, 1952)

$$Q = F\,\Delta x C_A \left[\frac{D_t}{\Delta x^2} - \frac{1}{6} - \frac{2}{\pi_n^2}\sum\frac{(-1)^n}{n^2}\exp\frac{(-D_n^2 n^2)}{x^2}\right] \tag{3.93}$$

As time increases and as the stationary state is approached, the summation term can be neglected and (3.93) simplified to

$$Q = \frac{DC_A F}{\Delta x}\left(t - \frac{\Delta x^2}{6D}\right) \tag{3.94}$$

The diffusion coefficient can also be determined from the time required to reach a steady state, the so-called "time-lag" (Fig. 3.12). This is done by extrapolating the linear part of the curve in Fig. 3.12 to $Q = 0$ when the intercept on the time axis gives the value of Δt, from which the diffusion coefficient can be evaluated (Barrer, 1941; Daynes, 1920). According to (3.94)

$$D = \frac{\Delta x^2}{6\Delta t} \tag{3.95}$$

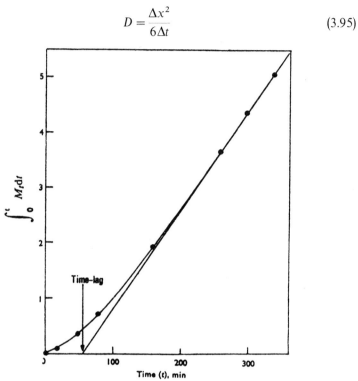

Fig. 3.12. Diffusion of Duranol Red 2B through polyethyleneterephtalate film $Q = \int Mdt = $ uptake of dye. (Reproduced with permission from McGregor, Peters and Ramachandran, 1968.)

B Non steady state diffusion

In the majority of cases with practical importance in dyeing, the diffusion process does not reach a stationary state. By definition the non-steady process is one in which

$$\left(\frac{\partial C}{\partial t}\right)_x \neq 0 \tag{3.96}$$

and therefore the concentration and its gradient change with time at every point in space, for the distribution of the dye in the substrate changes with time.

The first significant study of non-steady state diffusion of dyes was carried out in 1933. Neale and Stringfellow (1933) studied the rate of absorption of Sky Blue FF in cellophane.

About the same time Boulton *et al.* (1933) also studied the kinetics of dyeing uptake in viscose sheets. In this early work the substrate was immersed in dye solution at a given temperature, removed after certain time and analysed for

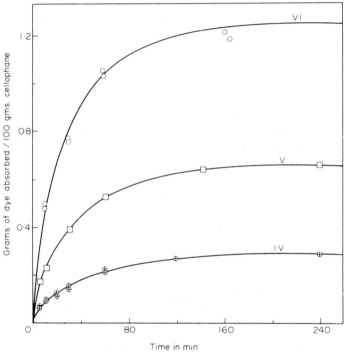

Fig. 3.13. Dye uptake by cellophane slabs. Curves calculated by equation (3.97). Points experimental results. Dye concentration 0·05 g/l; IV = NaA 2·0 g/l; V, NaA = 5 g/l; VI, NaA 12 g/l (A = acetate). (Reproduced with permission from Neale and Stringfellow, 1933).

dye uptake. The dye uptake was then plotted against time and analysed by means of Hill's equation (Hill, 1928) which was derived for calculating the dye uptake into a plain sheet where the penetration occurs from both sides

$$\frac{M}{M_\infty} = 1 - \frac{8}{\pi^2}\left[\exp\left(-\frac{D\pi^2 t}{4b^2}\right) + \frac{1}{9}\exp\left(-\frac{9D\pi^2 t}{4b^2}\right) + ..\right] \qquad (3.97)$$

where $2b$ is the thickness of the sheet, M is the uptake at t time, M_∞ is the uptake at infinite time, and D is the diffusion coefficient.

Some experimental curves obtained by Neale and Stringfellow are reproduced in Fig. 3.13.

Later Garvie and Neale (1938) developed a more sophisticated method for studying dye diffusion. Using the same apparatus as shown in Fig. 3.11 they measured the rate of diffusion of dyes through an array of parallel sheets of cellophane. In these experiments, however, the time of dyeing was kept short in order not to allow the attainment of a steady state. The dye concentration in the sheets was then determined as a function of distance from the surface.

From these types of curves M, the total amount of dye taken up by the total array calculated using the relationship

$$M = \int_0^x \left(\frac{\partial C}{\partial x} \right)_t dx \tag{3.98}$$

The integration generally is performed graphically by measuring the area beneath the curves.

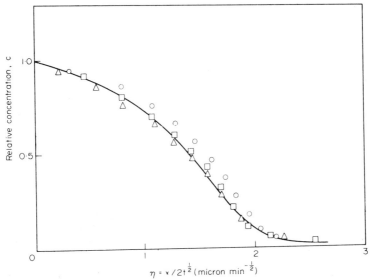

Fig. 3.14. Concentration distribution of Naphthalene Scarlet 4R in Nylon 6:6. x = distance from interface; t = time of dyeing; $\triangle$ = 25 min; $\square$ = 50 min; $\bigcirc$ = 82 min. (Reproduced with permission from McGregor, Peters and Petropolous, 1962.)

The diffusion coefficient can also be determined as a function of dye concentration. Firstly the value of the gradient $(\partial C/\partial x)$ can be obtained by constructing the tangent of C vs x curves at a given point x, then the diffusion coefficient calculated, at any dye concentration, by the equation

$$D(C) = -\frac{1}{2t} \frac{\partial x}{\partial C} \int_0^c x \, dC \tag{3.99}$$

C   Direct scanning of dye in fibres

Peters *et al.* (1959) and McGregor *et al.* (1961, 1962, 1965, 1968) developed a method for studying the diffusion of dyes. They used a sensitive micro-densitometer (Walker, 1955) which measured the dye intensity as a function of the distance in slabs of polymers.

Essentially the method involves the dyeing of a rectangular slab of the substrate for a given time and then sectioning it by means of a microtome. A

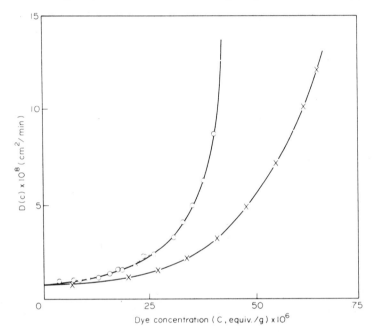

Fig. 3.15. Variation of diffusion coefficient of Naphthalene Scarlet 4R in Nylon 6:6 with dye concentration. Upper curve pH 6 and 3·2; lower curve pH 1·95. (Reproduced with permission from McGregor, Peters and Petropolous, 1962.)

slice normal to the dyeing direction is then cut from the slab and the dye content measured as a function of the distance from the original dye bath substrate interface using a specially modified microdensitometer. The dye concentration is determined by measuring the intensity of a thin light pencil shining through the substrate slice.

The substrate is gradually scanned by moving it perpendicular to the light beam and the amount of the dye taken up as a function of penetration is measured by means of a photocell. The results can then be represented as curves of dye concentration vs distance at various time intervals (Fig. 3.14) and the value of the diffusion coefficient measured as a function of dye concentration or time of dyeing.

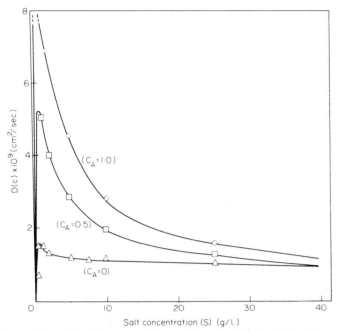

Fig. 3.16. Variation of the diffusion coefficient of Chlorazol Sky Blue FF in Cellulose with salt concentration: C_s—surface dye concentration. (Reproduced with permission from McGregor, Peters and Petropolous, 1962.)

McGregor *et al.* have studied the diffusion of a variety of dyes into cellulose, Nylon 6:6, polyesters and polyacrylonitryls. The technique proved very useful for determining the dependence of the diffusion coefficient on experimental conditions, e.g. dye concentration (Fig. 3.15), salt concentration (Fig. 3.16) and temperature (Fig. 3.17).

Another approach was that used by Blacker and Patterson (1969) to measure the dye distribution in cylindrical fibres. A magnified image of the cross-section of a dyed filament was projected onto a screen and then scanned by means of a narrow slit (Fig. 3.18). Assuming a radial symmetry, the dye distribution as the function of the distance from the centre of the fibre could be calculated from the scanned intensities. Good agreement was obtained between experiment and theory for nylon and polyester fibres (Fig. 3.19). Blacker's technique also lent itself to the determination of the molecular orientation of dyes in the fibre. The dichroism of the absorbed dye molecules could be measured using polarized light. The difference between the "fast" and the "slow" light beam was used for calculating the extent of orientation of the dye molecules in the fibre (Fig. 3.20).

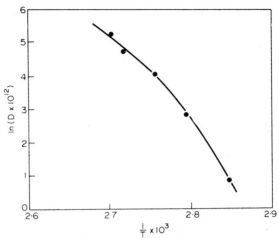

Fig. 3.17. The temperature dependence of the diffusion coefficient of Duranol Brilliant Blue CB in Courtelle film. (Reproduced with permission from McGregor, Peters and Ramachandran, 1968.)

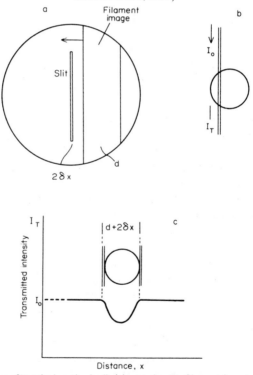

Fig. 3.18. Principles of "optical sectioning": (a) scanning the filament image with a narrow slit; (b) plan view showing the absorbing path within the filament measured by the slit; (c) effect of finite slit width on recorded trace. Transmitted intensity I_T plotted against distance x. (Reproduced with permission from Blacker and Patterson, 1969.)

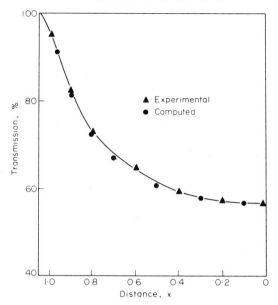

Fig. 3.19. Experimental and computed traces of light transmission against slit position for a filament uniformly penetrated with unorientated dye molecules. (Reproduced with permission from Blacker and Patterson, 1969.)

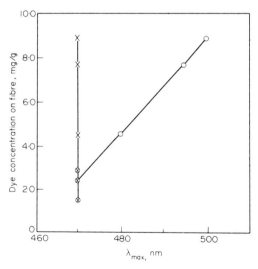

Fig. 3.20. Observed wavelength of maximum absorption of polarised light by filaments containing increasing concentrations of dichroic dye: C.I. Disperse Red 1 on nylon 6.6. O Polariser parallel to fibre axis   X Polariser perpendicular to fibre axis. (Reproduced with permission from Blacker and Patterson, 1969.)

D Measurement of the self diffusion coefficient

Two factors determine the rate of dye uptake, the affinity of the dye for the fibre and its migration velocity (Tyrrell, 1961). The experimental methods outlined in the previous sections measure the "chemical" or overall diffusion coefficient, the value of which depends on both the molecular migration velocity and on the thermodynamic interaction of the dye with the surrounding medium. The separation of these two factors is difficult.

The measurement of the self diffusion coefficient overcomes this difficulty. In these experiments the substrate is first dyed to a given equilibrium dye uptake level, then the dye fibre is transferred to a second "infinite" bath, containing the same dye in an identical concentration to that of the first bath. The dye molecules in the second bath are "labelled", i.e. one of their constituent atoms (generally, a H or a C atom) is a radioactive isotope.

An exchange of the radioactive dye molecules from the bath with the non-radioactive dye molecules adsorbed on the substrate will then occur. The rate of disappearance of the radioactive species from the bath into the fibre can be monitored by standard experimental techniques and the self or exchange diffusion coefficient, calculated from these experiments, can be interpreted as representing an average of the various molecular interactions in which the dye molecule is involved during its passage from the bath into the fibre. One of the commonly used averaging equations is the Stefan-Boltzmann equation, originally derived for the interpretation of gaseous diffusion. For the particular case of self diffusion it can be written in the form

$$\frac{1}{D_{obs.}} = \frac{X_1 + X^*_1}{D_{11}} + \frac{X_2}{D_{12}} + \frac{X_3}{D_{13}} + \dots \tag{3.100}$$

where X_1, X^*_1, X_2, X_3 respectively denote the mole fractions of the unlabelled and labelled dye of the solvent (component 2) of the polymer and of other components present in the system. The various diffusion coefficients D_{ij} represent the interactions between the dye molecule and the other components in the system.

A number of experimental investigations of diffusion in various polymeric materials was carried out using this technique (Chantrey and Rattee, 1969; Medley, 1957; White, 1968; Wright, 1954). Thus Chantrey and Rattee (1969) measured the diffusion of chloride ions in nylon, and concluded that the observed diffusion coefficient could be regarded as a harmonic average of two diffusion coefficients: one representing the migration of site-bound Cl^- from site to site, and the other reflecting on the migration of free Cl^- ions in water filled channels (Fig. 3.21).

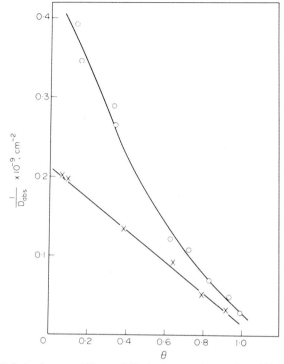

Fig. 3.21.   Relation between $1/D_{obs}$ and Cl, the degrees of saturation of binding site in nylon for Cl ions: O = constant ionic strength, x = varying ionic strength. (Reproduced with permission from Chantrey and Rattee, 1969.)

VI   THE   INTERPRETATION   OF   THE   DIFFUSION CONSTANT

To obtain a deeper insight into the mechanism of dye diffusion, the diffusion constant, obtained by analysis of the experimental measurements using the relationships outlined in the previous section, needs to be interpreted in terms of molecular processes. Two general approaches have been developed for the interpretation of the diffusion coefficient in molecular terms; the hydrodynamic theory and the kinetic or absolute rate theory. (For a detailed review see Tyrrell, 1961.)

A   Hydrodynamic theory

The approach was first suggested independently by Einstein (1906) and Sutherland (1905). Both authors regarded the diffusional flow as the result of a balance of two opposing forces; a driving force pushing the molecules ahead and a hydrodynamic resistance opposing the flow.

Einstein assumed that the driving force was the negative chemical potential gradient of the diffusing species. The chemical potential of component i can be expressed as a function of the concentration by

$$\mu_i = RT(\log C_i + \log f_i) + \mu_{0i} \tag{3.101}$$

where C_i denotes the concentration and f_i the activity coefficient. Therefore

$$\text{grad } \mu_i = RT \frac{d}{dC_i}(\log C_i + \log f_i)\,\text{grad } C_i \tag{3.102}$$

or for the one dimensional case when the flow proceeds in the direction of coordinate x

$$\frac{dC_i}{dx} = RT \frac{d}{dC_i}(\log C_i + \log f_i)\frac{dC_i}{dx}$$

$$= RT\left(\frac{1}{C_i} + \frac{d \log f_i}{dC_i}\right)\frac{dC_i}{dx}$$

$$= \frac{RT}{C_i}\left(1 + \frac{d \log f_i}{d \log C_i}\right)\frac{dC_i}{dx} \tag{3.103}$$

On the other hand, if the viscous resistance per molecule is represented by ξ, the frictional coefficient, then the total frictional force acting on one mole is

$$\text{Viscous resistance} = \xi N V_i \tag{3.104}$$

where V_i is the velocity of the diffusing substance and N Avogadro's number.

By relating (3.104) and (3.103), i.e. balancing the two forces, the velocity of the diffusional flow is

$$V_i = \frac{RT}{N\xi C_i}\left(1 + \frac{d \log f_i}{d \log C_i}\right)\frac{dC_i}{dx} \tag{3.105}$$

According to Fick's laws, the total flux can be expressed as the product of velocity and concentration of the species and can be set equal to the product of the diffusion coefficient and the gradient:

$$J_i = C_i V_i = -D\frac{dC_i}{dx} = -\frac{RT}{N\xi}\left(1 + \frac{d \log f_i}{d \log C_i}\right)\frac{dC_i}{dx} \tag{3.106}$$

and therefore from (3.105) and (3.106)

$$D = \frac{RT}{N\xi}\left(1 + \frac{d \log f_i}{d \log C_i}\right) \tag{3.107}$$

The value of ξ, the frictional coefficient of a sphere moving in a viscous fluid was calculated by Stokes and is given by

$$\xi = 6\pi\eta r \tag{3.108}$$

where η is the viscosity of the solvent and r the radius of the sphere.

Combination of (3.107) and (3.108) yields the Stokes-Einstein equation,

$$D = \frac{RT}{6\pi\eta rN}\left(1 + \frac{d\log f_i}{d\log C_i}\right) \qquad (3.109)$$

which is valid for a spherical molecule diffusing in a continuous homogeneous medium. For non spherical molecules the same type of relationship holds but instead of the factor six another numerical value has to be used.

Diffusion measurement can also be used to evaluate the shape of the molecule. This can be done by comparing a theoretical Stokes radius r_0 with the real Stokes radius r, computed by means of (3.108). We can calculate the value of r_0 using the relationship

$$r_0 = \left(\frac{3V}{4\pi N}\right)^{1/3} = \left(\frac{3M\bar{V}}{4\pi N}\right)^{1/3} \qquad (3.110)$$

where V is the molal volume, M the molecular weight and $\bar{V}$ the partial molal volume.

The ratio r/r_0, is known as the frictional ratio and can be correlated with the shape of the molecule, particularly for deviations from the perfect sphere. The values of the frictional ratios for ellipsoidal molecules with various axial ratios have been calculated theoretically (Herzog et al., 1933; Perrin, 1936; Svedberg and Pedersen 1940).

In deriving (3.108) we assumed that the dye molecule is in a very dilute solution and the dye molecule is surrounded by solvent molecules only. However, if the concentration of the solution is high, the dye molecule will encounter other dye molecules. Thus at finite dye concentrations, in addition to interactions between the solvent and dye molecules, interactions between two dye molecules have to be taken into account. The net result of these effects is a concentration dependence of the diffusion coefficient. Various equations have been suggested for the representation of the diffusion coefficient as a function of concentration. Thus Hartley and Crank (1949) derived the expression

$$D_d = \frac{RT}{N\eta(C_d + C_s)}\left(1 + \frac{d\log f_a}{d\log C_d}\right)(C_d D_d^0 + C_s D_s^0) \qquad (3.111)$$

where η is the viscosity of the solution, and C_s, C_d, D_s^0 and D_d^0 are the concentrations and the respective intrinsic diffusion coefficients of the dye in solvent and the solvent in dye at infinite dilutions. The Hartley and Crank equation assumes that the effective diffusion coefficient is a linear average of the two molecular intrinsic diffusion coefficients.

B Kinetic theory of diffusion

A different interpretation of the diffusion coefficient has been given by Eyring et al. (Glasstone et al., 1941). According to this theory, the medium in which the diffusion occurs (liquid or fibre) is an array of molecules in which occasional gaps occur spontaneously. The diffusing molecule jumps from one hole (A) into another (B) and during the transfer the molecule passes through an "activated" or transition state, T. From the transition state the molecule either falls back into hole A or goes ahead into B:

$$A \rightleftharpoons T \rightarrow B \tag{3.112}$$

Using a "quasi" thermodynamic argument, Eyring calculated the free energy of formation of the transition stage, ΔG^*. The diffusion coefficient can then be expressed in terms of the free energy of activation giving

$$D = \left(\frac{kT}{h}\right) \lambda^2 \exp\left(-\Delta G^*/RT\right) \tag{3.113}$$

$$= \left(\frac{kT}{h}\right) \lambda^2 \exp\left(-\Delta H^*/RT\right) \exp\left(\Delta S^*/R\right) \tag{3.114}$$

since according to thermodynamics

$$\Delta G^* = \Delta H^* - T \Delta S^* \tag{3.115}$$

The symbol k is the Boltzmann Constant, T the absolute temperature, h Plank's constant and λ the average distance of jump between two holes.

On the other hand the diffusion coefficient is a function of temperature and as such can also be expressed by the Arrhenius equation

$$\ln D = \ln A' - \frac{E}{RT} \tag{3.116}$$

where A' is called the frequency factor and E the activation energy.
Therefore a plot of $\log D$ vs $1/T$ should give a straight line, from the slope of which E, the energy of activation, can be calculated. The activation energy and the heat of formation of the activated state are interrelated (Tyrrell, 1961).

$$E = RT + \Delta H^* \tag{3.117}$$

The quantities ΔH^* and ΔS^* are measures of the heat and the entropy changes which accompany the diffusion process. Their quantitative interpretation in terms of molecular processes is difficult and requires the postulation of molecular models.

VII MODELS FOR MIGRATION OF DYES IN TEXTILE FABRICS

The theoretical treatments (Tyrrell, 1961) interpreting diffusion coefficients in terms of molecule quantities were developed for fluids (liquids and gases). Textile fibres are generally polymeric materials which contain a certain amount of absorbed liquid. Essentially two models have been formulated to explain the mechanisms of dye migration in them, notably the rigid porous matrix and the dynamic polymeric-chain model. The former is generally applicable to temperatures below the T_g (the glass transition temperature) whereas the latter is used to explain the migration of dyes above the T_g (Crank and Park, 1968; Duffy and Olson, 1972).

A   Porous matrix model

According to this model, the textile fibre can be regarded as a solidified sponge, (Davis and Shapiro 1964; Vickerstaff, 1954) a rigid matrix in which a maze of interconnected pores exist. The pores are filled with water and the dye enters from the bath liquid into the pores and penetrates the fibre by diffusing along these water filled channels. Dye binding sites are distributed on the surface of the pore walls. Dye molecules move along in the aqueous phase of the pore and will, as a consequence of thermal motion, collide with a binding site from time to time and become bound and therefore immobilized. However, depending on the strength of the binding, the dye molecule will desorp after a certain time, re-enter the aqueous phase and resume its movement towards the interior of the fibre. This model is based on three assumptions:

i.   the migration of the dye molecules only occurs in the water filled micro channels (pores)

ii.   a dynamic reversible equilibrium exists between the migrating dye molecules and the dye binding sites on the walls of the pores

iii. the dye molecules while adsorbed onto the sites are immobilized.

There are refinements of the model omitting assumption iii and substituting instead the assumption that a second migration process occurs on the walls (Medley, 1965).

B   Diffusion process in fibres with simultaneous adsorption

According to this model, a substantial proportion of the dye molecules that penetrate the fibre are adsorbed and immobilized, reducing the effective rate. When calculating the diffusion coefficient of a dye molecule from uptake data (i.e. M vs $\sqrt{t}$ plots) we have to correct the adsorbed fraction of the dye. Assuming that the rate of interaction of a dye molecule with the adsorption site is instantaneous (fast compared to the rate of diffusion), then the dye concentration at any given point in the pore will be in thermodynamic equilibrium with the adsorbed dye on the pore walls.

To proceed further we have to define the shape of the adsorption isotherm, i.e. the mathematical relationship between the concentration of the free dye in the aqueous phase and the amount of dye adsorbed on the pore surface. The simplest case is when the adsorption is linear, then

$$\frac{C_b}{C_f} = K \tag{3.118}$$

and a linear relationship exists between the C_f and C_b the concentrations of the free and bound dye respectively. The total dye concentration, C_T in the fibre is

$$C_T = C_f + C_b \tag{3.119}$$

but from (3.118)

$$C_b = C_f K$$

The concentration of free dye at any point is then

$$C_f = \frac{C_T}{1+K} \tag{3.120}$$

Since only the free dye will diffuse we can write the diffusion equation as

$$\frac{\partial C_f}{\partial t} = D \frac{\partial^2 C_f}{\partial x^2} = \frac{\partial C_T}{\partial t} = \frac{D}{1+K} \frac{\partial^2 C_T}{\partial x^2} \tag{3.121}$$

The effective diffusion coefficient therefore will be

$$D' = \frac{D}{1+K} \tag{3.122}$$

Equation (3.121) is identical with the simple diffusion equation, provided we substitute the diffusion coefficient by D' as in (3.122).

It follows that the larger the adsorption (larger K) the slower will be the rate of dyeing. If the adsorption isotherm is non-linear, the isotherm equation can be expressed by the general equation

$$C_b = K'C_f^n \tag{3.123}$$

and the problem becomes much more complicated. The diffusion equation will become non-linear and thus not accessible to an analytical solution. This problem has been discussed in detail by Crank (1956).

In all the latter cases, i.e. non-linear adsorption isotherms, the apparent diffusion coefficient is concentration dependent. A number of techniques are available for the evaluation of the concentration dependence of the diffusion coefficient from experimental data (Fujita, 1961). These methods generally involve the numerical determination of the integral diffusion coefficient $\bar{D}$ defined as

$$\bar{D} = \frac{1}{C_1 - C_2} \int_{C_1}^{C_2} D(C) \, dC \tag{3.124}$$

where C_1 and C_2 are the concentration limits. Using various mathematical techniques we can then evaluate the differential diffusion coefficient as a function of C.

C Evaluation of the diffusion data in terms of the pore model

Once D as a function of C has been obtained we can proceed to evaluate the reasons for the concentration dependence of D and obtain valuable information concerning the mechanism of dyeing. Thus for instance, the nature of the adsorption isotherm can be calculated from the dependence of the value of D on C.

The relationship between the diffusion coefficient and a, the thermodynamic activity is given by

$$D = D_0 \frac{d \log a}{d \log C} = \frac{RT}{N\xi}\left(1 + \frac{d \log f}{d \log C}\right) \tag{3.125}$$

where f is the activity coefficient and C the dye concentration and

$$D_0 = \frac{RT}{N\xi} \tag{3.126}$$

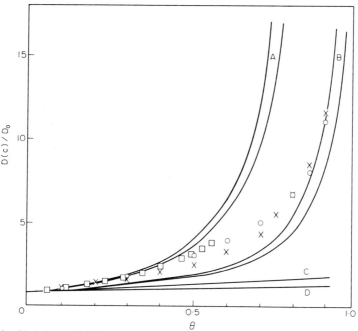

Fig. 3.22. Variation of D (c)/Do with θ, the degree of saturation for Naphthalene Scarlet 4R in Nylon 6:6. $\square$ = pH 6·6; $\bigcirc$ = pH 3·2, 80°C; x = pH 3·12, 60°C. Curve A calculated on the assumption that D(c) $1/1-\theta$; Curve B on the assumption that D(c) $(1/1-\theta)^2$. Equations C and D assume different model. (Reproduced with permission from McGregor et al., 1962.)

The explicit form of the concentration dependence of D on C depends on the assumption made regarding the functional relationship between a and C. This problem has been discussed in some detail (Atherton *et al.*, 1955, 1956).

If the driving force for the diffusion is regarded as being gradient of the chemical potential of the dye, then for the diffusion coefficient one obtains the expressions

$$D = D_0 \cdot {}^1/(1-\theta) \tag{3.127}$$

where

$$\theta = \frac{\text{occupied binding sites}}{\text{total number of dye binding sites}} \tag{3.128}$$

If, on the other hand, the activity gradient is regarded as the driving force then D can be expressed as

$$D = D_0 \frac{1}{(1-\theta)^2} \tag{3.129}$$

Experimental results (McGregor *et al.*, 1962) appear to be more in agreement with (3.127) then with (3.129). (Fig. 3.22).

D   Diffusion coefficients of ionic dyes (acid and basic)

When acid or basic dyes diffuse into a fibre which contains ionic groups of opposite sign as the binding site, the relationships become more complicated; additional constraints (e.g. the requirement of preserving electroneutrality in every part of the solution and local electrical potentials) come into consideration. This is the case when wool, nylon or acrylate fibres are dyed. A number of authors have treated the problem. Recently J. A. Medley (1964), who studied the dyeing of wool by acid dyes, formulated the following theory.

The dye ion and its counterion (e.g. Acid Orange C.I. anion and H^+) have different migration velocities or mobilities. Since electroneutrality has to be preserved at every point of the fibre, the difference of the mobilities of the two ion species causes the slow ion (dye anion in the case of acid dye) to accelerate.

Medley derived the following equation for ionic dyes, based on the Crank–Hartley equation,

$$D(C) = \frac{RT}{F^2} \frac{\Lambda_H \Lambda_A}{\Lambda_H + \Lambda_A} \frac{Z_H + Z_A}{Z_H Z_A} \left(1 + \frac{d \log f \pm}{d \log C}\right) \tag{3.130}$$

where F is the Faraday constant, 96 500 coulombs, Λ_H and Λ_A, are equivalent limiting conductivities of the dye anion and the cation (generally H^+), and Z_H and Z_A denote the valencies of the cation and acid dye anions respectively.

We assume the mean activity coefficient to be the geometric mean of the individual ionic activity coefficient

$$f\pm = \sqrt{f + f -} \tag{3.131}$$

On the other hand the self diffusion coefficient of the anion alone, measured by radio tracer technique at a constant cation concentration, is

$$D_A^* = \frac{RT}{Z_A F} \Lambda_A \tag{3.132}$$

The transport number of the cation is

$$t_H = \frac{\Lambda_H}{\Lambda_A + \Lambda_H} \tag{3.133}$$

(3.130) can therefore be written as

$$D(C) = D^* (Z_A + t) \left(1 + \frac{d \log f \pm}{d \log C} \right) \tag{3.134}$$

Three theories are available for describing the binding of ionic dyes to keratin (Breuer, 1964; Gilbert and Rideal, 1944; Peters and Speakman, 1949). Using either of these theories we can derive an expression for the value of $d \log f \pm / d \log C$.

Medley, using the Gilbert–Rideal theory (Gilbert and Rideal, 1944), calculated for monosulphonic dyes the ratio of $D(C)/D^*$ as a function of dye concentration and compared it to experimentally measured ratios. A fairly good agreement was obtained suggesting that (3.134) describes the variation of diffusion coefficient of electrolytic dyes fairly well (Fig. 3.23).

E Applicability of the pore model

In order to test the applicability of the pore model to a given type of fibre, we must determine and compare the ratio of the diffusion coefficients of various dyes in water and in the fibre. If these ratios are constants for a series of dyes, the pore model can be regarded as valid. The ratio of the diffusion coefficients in the fibre and in water is the tortuosity factor ρ and is a measure of the volume fraction of the fibre occupied by pores. The measured activation energy calculated from the temperature dependence of D' by the Arrhenius equation is the sum of two terms: the activation of diffusion and the heat of adsorption of the dye

$$\ln D' = \ln D - \ln (1 + K) \tag{3.135}$$

and

$$\frac{d \ln D'}{d \, 1/T} = \frac{(E - \Delta H)}{RT} \tag{3.136}$$

where

$$\frac{d \ln (1 + K)}{d(1/T)} = \frac{-\Delta H}{RT} \tag{3.137}$$

Should the Arrhenius equation not hold or the value of E be very different from that of the dye in water, we can assume that the dye migration follows a different mechanism of diffusion and that the pore model is not applicable.

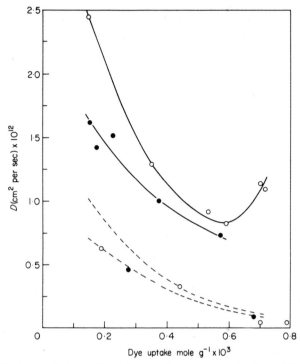

Fig. 3.23. Diffusion coefficients of naphthalene Orange G as a function of dye uptake with fibre. Experimental points ◯ wool; ● Horn. Solid curve measured values, broken curves calculated for D* by equation 3.134. (Reproduced with permission from Medley, 1964.)

VIII   THE FREE VOLUME MODEL OF DIFFUSION

Below a certain characteristic temperature, denoted as T_g the glass transition temperature, completely amorphous polymeric materials behave as rigid bodies or glasses. Above the T_g polymeric material becomes rubbery and behaves as highly viscous fluid. We can use two physical quantities to characterize the mechanical and rheological properties of polymers; the elastic modulus ε' and the storage or viscous modulus ε''. The first measures the elasticity of the polymer, the force required for a unit deformation, whereas the latter gives a measure of the energy dissipated (not recovered) during a cycle of extension and relaxation. The larger the value of ε'', the more energy is dissipated. Another useful quantity for characterizing the rheological properties of polymers is the loss tangent defined by $\tan \delta = \varepsilon''/\varepsilon'$.

By plotting the value of ε' and $\tan \delta$ against temperature (Murayama et al., 1968) the following typical curves are obtained (Fig. 3.24).

Below the T_g the value of ε' is high and the polymer behaves as a rigid glass. In this region dyes can only move through the pores and cracks in the polymer. Above the T_g the polymeric chains acquire a large degree of mobility, the value of ε' drops and the polymeric material behaves essentially as a highly viscous liquid. The whole rigid fabric of polymeric material disappears and it becomes a rubbery viscous uniform material. There are no more solvent-filled pores, but only a highly viscous mixture of polymer and solvent.

Above the T_g the migration of dyes is linked to the amount of free volume available (the volume which is not occupied by the chains of the polymer

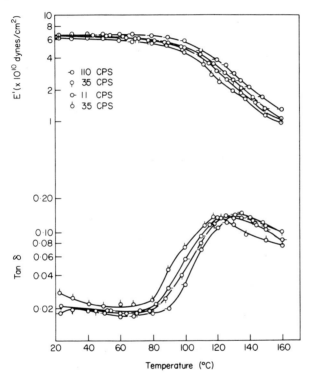

Fig. 3.24. Dynamic modulus E' and loss tangent δ as a function of temperature for polyethylene fibre phtalate at four different frequencies as indicated. (Reproduced with permission from Murayama *et al.*, 1968.)

material itself). The larger the free volume, the greater is the probability that suitable voids will open up periodically through which the dye molecules move.

The viscosity of a polymer melt and the transport of dye molecules through polymeric networks are governed by the same molecular process, notably the frequent creation of large enough voids between polymeric chains. The bulk viscosity of the polymer essentially depends upon the number of voids available for the accommodation of segments of neighbouring polymeric chains, thus allowing a movement of the chains. Similarly the rate of dye transport depends upon the frequency of formation of holes which are large enough to accommodate a dye molecule. Therefore, to calculate the rate of dye transport we have to be able to predict the formation of holes larger than a given critical volume.

Bueche (1956, 1959, 1962) has developed a theory for predicting the probability of hole formation based on simple statistical mechanics. The number of holes larger than critical volume V^* is given by

$$n_b = \frac{n \int_{b*}^{\infty} \exp\left(-4\pi\sigma b^2/kT\right) \mathrm{d}b}{\int_0^{\infty} \exp\left(-4\pi\sigma b^2/kT\right) \mathrm{d}b} \tag{3.138}$$

where σ is the specific surface energy of the polymer, b the radius of the hole and b^* the radius of the hole with volume V^*.

Integration of (3.138) yields

$$n_b = n\left(\pi\sigma^*/kT\right)^{1/2} \exp\left(-\sigma^*/kT\right) \tag{3.139}$$

where
$$\sigma^* = 4\pi\sigma b^{*2} \tag{3.140}$$

Bueche's theory did not take into account "cooperative" hole formation, i.e. the situation when two small holes, neither of which individually is large enough to accommodate a dye molecule, form adjacent to each other. Bueche's theory would count this two adjacent hole formation as too small for transport in spite of the fact that collectively the two adjacent holes might be large enough to accommodate a dye molecule. This short coming was corrected by Cohen and Turnbull's theory (Cohen, 1959 and Turnbull, 1961) which also provided a molecular basis for the empirical Doolittle equation (Doolittle and Doolittle, 1957) representing the bulk viscosity of a polymeric melt as a function of its free volume

$$\ln \eta = \ln A + B\left(V - V_f\right)/V_f \tag{3.141}$$

where V is the total volume and V_f the free volume.

The quantities A and B are numerical constants which can be expressed in terms of molecular quantities by means of the Cohen–Turnbull theory.

A   Factors affecting the free volume of polymers

Since the dye diffusion through polymeric materials depends strongly on the available free volume in the polymeric fibre, it is important to know the various factors that influence the amount of free volume present. The specific volume of a polymer is a linear function of the reciprocal of the number average molecular weight, M_n. Fox and Flory (1950) proposed that the same type of relationship applies to the free volume dependence. The phenomenon can be explained by assuming that the terminal segments of a polymeric chain are more bulky than those in the middle of the chain, and consequently, the end segments will pack less tightly than the other residues of the polymer and will make a larger than average contribution to the free volume of the melt.

The molecular weight dependence of V_f at a given temperature can therefore be expressed as

$$V_f = V_{f,\infty} - \frac{K}{M}$$ (3.142)

where

$$K = 2\rho N V_f{}'/\alpha$$ (3.143)

in which ρ, is the density of the polymer, N Avogadro's number, $V_f{}'$ the free volume per terminal segment and α the bulk-thermal volume expansion coefficient. Furthermore it can be shown that for most of the polymeric materials the following equation holds in good approximation

$$V_{f,g} = 0.025 V_t$$ (3.144)

where $V_{f,g}$ is the free volume of the polymer at the glass transition temperature and V_t the total volume. Therefore a similar relationship to the one expressed in (3.142) will also hold for the molecular weight dependence of T_g, the glass transition temperature,

$$T_g = T_{g,\infty} - \frac{K}{M}$$ (3.145)

where $T_{g,\infty}$ is the limiting glass transition temperature for a polymer with infinite molecular weight. The validity of (3.145) has been tested experimentally.

B   Effect of diluent on the free volume and the glass transition temperature

The presence of a small molecular weight component with a polymer changes its T_g and, consequently, also influences the available free volume. The change in T_g can be explained by assuming that the free volume of the polymer and of the diluent are, in a first approximation, additive quantities (Braun and Kovacs, 1965):

$$f = Xf_{f,p} + f_{f,dil}(1-X)$$ (3.146)

where f is the fraction of the free volume in the polymer diluent mixture $(V_{f,m}/V), f_p$ is the fraction of the free volume in the polymer $(V_{f,p}/V), f_d$ is the fraction of the free volume in the diluent $(V_{f,d}), V$ is the total volume and X is the volume fraction of the polymer in the mixture.

The glass transition temperature as a function of the volume fraction of the polymer in a polymer–diluent system is then given by

$$T_g = \frac{X\alpha_p T_{g,p} + (1-X)\alpha_g T_{g,d}}{X\alpha_p + (1-X)\alpha_d}$$ (3.147)

where α_p is the thermal expansion coefficient of the polymer, α_d is the thermal expansion coefficient of the diluent, T_g is the glass transition temperature of the mixture, $T_{g,p}$ is the glass transition temperature of the polymer and $T_{g,d}$ is the glass transition temperature of the diluent.

At high diluent concentrations the simple additivity of the free volumes does not hold and we require a more complicated relationship. This accounts for the non-linear relationship observed (Jenckel et al., 1953) for T_g at high volume fraction of diluents (Fig. 3.25).

C   Temperature dependence of V_f

The fraction of available free volume depends on the temperature. A simple relationship is generally sufficient to describe the temperature dependence of the volume expansion. It can be assumed that the increase in the total volume with temperature is entirely due to the increase of the free volume:

$$V_f(T) = V_{f,g} + \alpha(T - T_g) \tag{3.148}$$

where $V_{f,g}$ is the free volume of the glass transition temperature and α the bulk-thermal volume expansion coefficient.

Williams, Landel and Ferry (1955) have derived an empirical relationship for the temperature dependence of the bulk viscosity of polymeric materials above their glass transition temperatures (the WLF equation):

$$\log \frac{\eta_1 \, \rho_1 \, T_1}{\eta_2 \, \rho_2 \, T_2} = \frac{f_1 \, (T_2 - T_1)}{f_2/\alpha + T_2 - T_1} \tag{3.149}$$

where η denotes the viscosity, ρ the density and f the fraction of free volume of the polymer at the respective temperature indicated by the subscript.

The WLF equation, together with (3.148) yields a relationship between the free volume, the density, the viscosity and the temperature of polymeric materials above their T_g.

D   Free volume theory of dye migration of polymers

The pore model for dyeing of textile fibres did not adequately explain the dyeing kinetics of many synthetic fibres (polyesters, acrylates) and therefore another approach was required. The need for this second approach was highlighted by Rosenbaum's extensive work (Rosenbaum, 1963a, 1963b, 1964, 1965a, 1965b, Rosenbaum and Goodwin, 1965) on the kinetics of dyeing of acrylic fibres with cationic dyes. Plotting $\log D$ of Malachite Green in polyacrylonitrile fibres against $1/T$ Rosenbaum obtained a sigmoid curve instead of straight lines which would have been expected on the basis of the pore theory (Fig. 3.26). Furthermore, the activation energy is not a constant but changes markedly with temperature (see inset in Fig. 3.26).

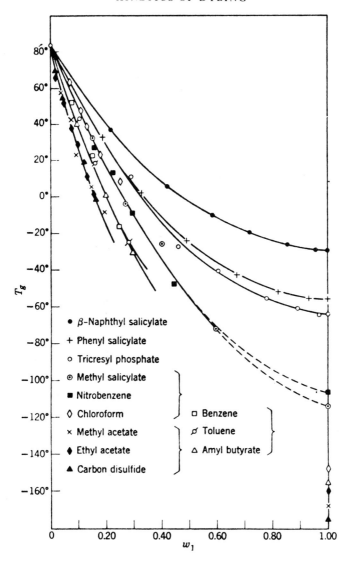

Fig. 3.25. Glass transition temperature of polystyrene as a function of ω_1 the mole fraction of diluent. (Reproduced with permission from Jenckel *et al.*, 1953.)

The free volume theory of polymers offers a suitable approach for the explanation of these results (Rosenbaum, 1965a, 1965b). The migration mobility m of a small molecule in a polymeric melt is an exponential function of V_f the free volume (Fujita *et al.*, 1960)

$$m = A \exp\left(-B/V_f\right) \qquad (3.150)$$

where A and B represent numerical constants characteristics for the specific dye-polymer system.

According to Einstein's equation the diffusion coefficient of a molecule is

$$D = RTm = \frac{RT}{\rho} \qquad (3.151)$$

where ρ is the frictional coefficient of the molecule, therefore

$$D = RTA \exp\left(-B/V_f\right) \qquad (3.152)$$

The available free volume in a polymer–dye (diluent) system depends on both the temperature and on the concentration of dye (diluent) (see (3.145)

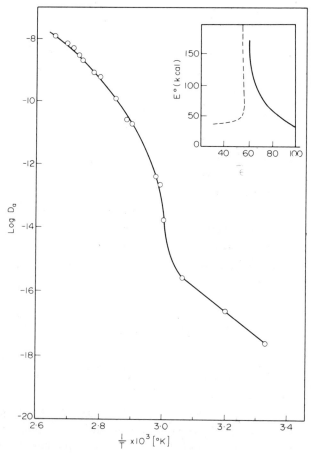

Fig. 3.26. Arrhenius plot of apparent diffusion coefficient D_a of Malachite Green in experimental polyacrylonitrile fibre. Inset plot shows variation with temperature of apparent activation energy. (Reproduced with permission from Rosenbaum, 1965.)

and (3.147)). Thus (3.152) can be further developed to give D as a function of these quantities.

The free volume theory of polymers (Rosenbaum and Goodwin, 1965) predicts the existence of a simple relationship between the diffusion coefficient of a dye and the bulk viscosity of the polymer

$$\ln\left(\frac{D}{RT}\right) = C - B\ln\eta \tag{3.153}$$

where C and B are constants.

Instead of substituting explicit functions for V_f which express the concentration and temperature dependences of the free volume, it is simpler for practical purposes to express the dye diffusion coefficient in terms of quantities characterising the polymer viscoelastic properties.

Bell (1968) has examined in some detail the validity of this assumption, by measuring the diffusion coefficient of the dyes Kiton Red 2G and Edicol Supra Geramine 2GS in Nylon, and correlating them with ε'', the dynamic loss modulus, measured independently. Bell obtained good correlations with the dynamic loss modulus (Fig. 3.27). It is often more useful to plot the quantity $\varepsilon''/_\omega$ where ω denotes the circular frequency at which ε'' is measured.

IX   THE DYEING OF SEMICRYSTALLINE POLYMERIC FIBRES

Many of the fibres used in textiles are drawn during their manufacture. This induces considerable crystallinity in these fibres which further complicates the dyeing characteristic. In order to understand the way crystallization influences dyeing characteristics we have to discuss the factors governing the crystallinity of polymer and the rate of crystal formation.

A   Crystallinity of the polymers

We can measure the crystallinity of polymers by a number of ways, one being X-ray diffraction in a Debye camera (Mandelkern, 1964). When the fibre is heated, the crystalline X-ray powder reflections gradually disappear and the polymer becomes amorphous. The temperature at which the crystallinity completely disappears is known as T_m the melting point of the polymer; its value is determined by the ratio of heat to entropy of crystallization

$$T_{m,0} = \frac{\Delta H_m}{\Delta S_m} \tag{3.154}$$

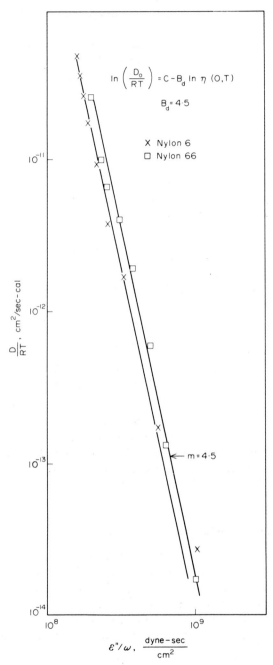

Fig. 3.27. Relation between dynamic loss modulus and dyeing rate. (Reproduced with permission from Bell, 1968.)

where ΔH_m and ΔS_m are the molal enthalpy and entropy differences between the polymer molecules in their crystalline and molten state.

Stretching of the fibre affects its crystallinity. As stretching induces a higher degree of order, the partial entropy of an amorphous polymeric chain in a strained fibre is lower than in an unstrained fibre. Therefore the melting point will increase. The entropy of the crystalline regions will be unaffected by the strain, as will be the partial molal heats of the two conformations. Flory (1953) calculated the effect of stretching in the melting point of polymers and obtained the equation

$$1/T_m - 1/T_{m,0} = (R/\Delta H_m)(6/\pi n)^{1/2} - (\lambda^2/2 + 1/\lambda)/n \qquad (3.155)$$

where λ is the extension ratio and n is the number of monomeric units between cross links or chain entanglements. An example for the effect of stretching on crystallinity is given in Fig. 3.28. Unlike in simple substances, in polymers the

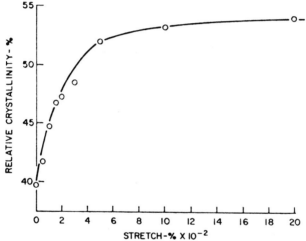

Fig. 3.28   Effect of stretch on crystallinity of saponified acetate. (Reproduced with permission from Sprague, 1967.)

melting point is not a very sharply defined temperature. The reason for this is that in polymers the crystalline regions are small and therefore have a substantial surface energy E_s. Therefore for semicrystalline polymers (3.154) should really be written as

$$T'_m = \frac{(\Delta H_m - E_s)}{\Delta S_m} \qquad (3.156)$$

The quantity E_s will depend on the degree of crystallinity and the distribution of crystallite sizes.

The presence of a diluent also affects the value of T_m. The diluent generally only penetrates the amorphous region of the polymers whereas leaves the crystallites intact. Interaction with the polymeric chains lowers the chemical potential of the amorphous regions without affecting the crystalline areas (no penetration). The value of ΔH_m is therefore diminished and T_m depressed as the presence of diluent hardly affects ΔS_m. The functional relationship between polymer T_m as a function of diluent content was first derived by Flory (1953)

$$1/T_m = \frac{1}{T_{m,0}} + R\left[\bar{V}/(\bar{V}\Delta H_m)\right]\left[V_1 - \chi V_1^2\right] \qquad (3.157)$$

where $\bar{V}$ and $\bar{V}$ are the molar volumes of the polymer repeat unit and of the diluent, V_1 is the volume fraction of the diluent and χ is the polymer–solvent interaction parameter.

The rate of crystallization will depend on the rates of two processes: the rate of nucleation and the rate of crystal growth.

The rate of nucleation will be practically zero at the upper limit of the

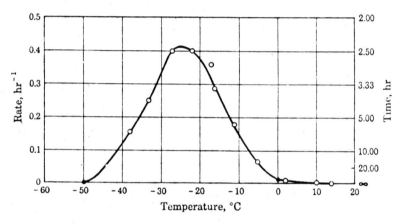

Fig. 3.29 Plot of the rate of crystallization of natural rubber over an extended temperature range. The rate plotted is the reciprocal of the time required for one-half the total volume change. (Reproduced with permission from Wood and Bekkedahl, 1946.)

melting point range. Any nucleus formed at this temperature will redissolve immediately owing to the large value of the surface energy term. At lower temperatures the free energy becomes negative since

$$\Delta G = \Delta H - T\Delta S + E < 0 \qquad (3.158)$$

Therefore the crystal nuclei formed will remain stable. As the temperature is further lowered, the growth process slows down owing to an increased viscosity of the melt and lower diffusibility of the polymeric chains. The dependence of the rate of crystallization on temperature (Mandelkern, 1964) therefore exhibits a maximum (Fig. 3.29).

B Effect of crystallinity on dye diffusion

Sprague (1967) demonstrated that pre-stretching of the fibres has a major effect on the rate of dyeing of acetate rayon with Disperse Blue 3 (Fig. 3.30).

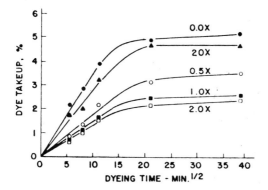

Fig. 3.30 Dye uptake at various degrees of crystallinity (X). (Reproduced with permission from Sprague, 1967.)

The relationship between dye diffusion and draw ratio was not straightforward, however. The dyeing rate goes through a minimum as the draw ratio increases. Dumbleton et al., (1968) has carried out a detailed study on the correlation between the crystallinity and the dye diffusion in Nylon and polyester fibres.

The T_g of polymeric fibres depends on the total amount of crystallinity and on the average size of the crystallites. The glass transition temperature has been shown (Dumbleton and Murayama, 1967) to be a linear function of the quantity X'/V' (where X' is the total amount of crystallinity and V' the average volume of a crystallite (Fig. 3.31).

As we have shown in the previous section, the rate of dyeing depends on the existence of a sufficient number of holes between the polymer chains which are large enough to accommodate dye molecules, and that a direct relationship exists between the available number of holes, i.e. the free volume, and the T_g. Consequently the ratio of X'/V', by influencing T_g, will also affect the diffusion coefficient. Dumbleton et al. (1968) measured the effect of temperature on both D and X'/V' in nylon and polyester fibres and found that the dye diffusion coefficients were lower in the highly drawn samples which exhibited high degrees of crystallinity. The dye diffusion coefficient also depends in a complicated way on the annealing temperature (Fig. 3.32). Both the temperature and drawing ratios affected the value of X' and of V', but to different degrees, and high drawing ratios increased the value of T_m owing to the decrease in the entropy of the amorphous phase lowering the value of X'. On the other hand the average crystallite size becomes less, owing to the effect of drawing on the kinetics of crystallization.

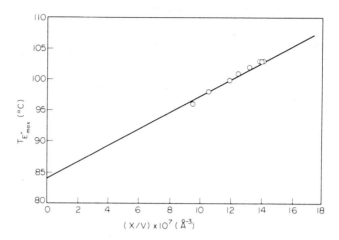

Fig. 3.31   Plot of the $T_{E''}$ glass transition temperature measured as the maximum of the storage modulus vs temp. curve as a function of the ratio X'/V'. (Reproduced from Murayama *et al.*, 1968.)

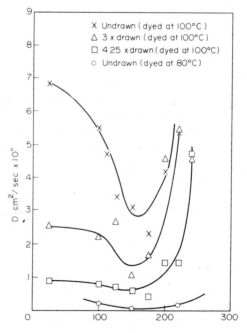

Fig. 3.32   Dye diffusion coefficient as a function of annealing temperature. (Reproduced with permission from Dumbleton *et al.*, 1968.)

X   DYE TRANSPORT FROM MOVING LIQUID

We assumed in the treatment of the previous sections that the dye transport from the bath into the substance occurs under static conditions. In practice, this is not the case however. Industrial dyeing equipment, whether designed for continuous or for batch operation, generally incorporates mechanical means of agitation or of stirring to facilitate dyeing. The diffusion of the dye from the bath into the cloth or fibres therefore occurs under conditions where the substrate and the dye-containing liquid are in motion relative to each other (convective diffusion). The rate of transport across the dye bath–substrate interface depends on the relative velocity of motion of these two phases. It is possible to discuss the problem in terms of a simplified model: diffusion from a fluid stream flowing past a solid plate.

The motion of volume element of an incompressible fluid can be described by two differential equations (Levich, 1962): the continuity equation expressing in mathematical terms the law of conservation of matter,

$$\text{div } V = \frac{\partial V}{\partial x^x} + \frac{\partial V}{\partial y^y} + \frac{\partial V}{\partial z^z} = 0 \tag{3.159}$$

where V is the velocity of the fluid and V_x, V_y and V_z represent the vectorial components of the velocity in the three dimensional space; and the Navier–Stokes equation

$$\rho \left(\frac{\mathrm{d}V}{\mathrm{d}t} \right) = - \text{grad } p + \eta V + f \tag{3.160}$$

where ρ is the density of the fluid, grad p the pressure gradient, η the viscosity and f the vector of external forces acting on the fluid (e.g. gravitation).

The exact solution of these differential equations, i.e. the distribution of the velocities, will depend on the conditions of the fluid motion. Two types of fluid flows are possible—laminar flow and turbulent flow.

A   Laminar flow

In laminar flow we can regard the liquid as being composed of a series of layers which, under the influence of an external force field (Fig. 3.33) slide on top of each other. The first layer, next to the solid surface, remains static with respect to the solid surface. The velocity V_x, a fluid layer in the direction of flow x, can be expressed as a function of x, the distance from the solid liquid interface:

$$V_x \approx \frac{U_{0,y}}{\delta_0} \tag{3.161}$$

where $U_{0,y}$ is the velocity of the main stream, the fluid flow very far from the

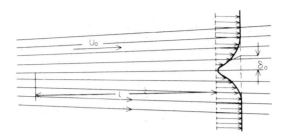

Fig. 3.33   Laminar flow past a plate. (Reproduced with permission from Levich, 1962.)

surface and δ_0 is the thickness of the boundary layer. The latter is defined as the distance from the surface of the plate to the point where V_x attains a value of 90% to that of U_0.

If the length of the plate along which the fluid flow is l, the thickness of the boundary layer is approximately given by

$$\delta_0 \approx \sqrt{\frac{vl}{U_0}} = \sqrt{\frac{l}{R_e}} \tag{3.162}$$

where, v is the kinetic viscosity,

$$v = \frac{\eta}{\rho} \quad (\rho \text{ density of the flow})$$

and the quantity R_e is known as the Reynolds number which is expressed by

$$R_e = \frac{U_0 l}{v} \tag{3.163}$$

The velocity distribution of the fluid in the viscosity of the solid surface is represented by the curve shown in Fig. 3.33.

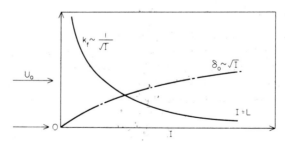

Fig. 3.34   Boundary layer thickness and shear stress as a function of X, the distance along the plate. (Reproduced with permission from Levich, 1962.)

B Turbulent flow

At high values of the Reynolds number the steady laminar flow gives a chaotic motion in the liquid, allowing the definition of only an average net flow in a given direction. Transition from laminar to turbulant flow occurs when and where the Reynolds number exceeds a critical value, R_{e_c}.

When the Reynolds number exceeds its critical value, fluctuations which occur spontaneously around the steady state (i.e. the laminar flow) are not dampèd down any more, but began to grow to major disturbances. The conditions under which this situation occurs have recently been examined in detail and defined by Glansdorff et al. (1971). For practical considerations it is sufficient to note that a fluid flows where the Reynolds number exceeds the value of about 1500, turbulant flow generally sets in. However, the exact value of R_{e_c} will depend on the prevailing condition, and in particular on the geometry of the system. If the flow of a fluid stream is next to an infinite plate, the fluid layer in the immediate proximity of the solid surface can be regarded as being stationary; moving away from the surface the flow becomes laminary until at a certain thickness of the fluid layer the Reynolds number exceeds its critical value and turbulant flow starts (Fig. 3.35).

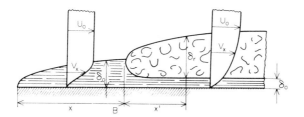

Fig. 3.35 Schematic representation of the formation of turbulent boundary layer on the plate. (Reproduced with permission from Levich, 1962.)

C Convective diffusion

The transfer of a dye from a dye bath into the fibres or cloth occurs in two steps; the transfer of the dye to the surface of substrate, and the penetration of the dye into the substrate. The rate of dyeing will be determined by the slowest process in this transport chain.

Dyeing can be regarded as heterogeneous reaction and can be described by the appropriate mathematical treatments. Nernst (1904) considered the kinetics of diffusion controlled heterogeneous reactions and proposed the "boundary layer model" for explaining the experimental results. Essentially this model assumes that the rate of heterogeneous reactions is governed by the rate of diffusion across a liquid layer adjacent to the solid surface:

$$Q = D \frac{(C_s - C_0)}{\delta} S \qquad (3.164)$$

where, C_s is the concentration of dye at the substrate surface, C_0 the concentration of dye in the bulk of the solutions, δ the thickness of the liquid layer next to the surface, S the surface area of the substrate and Q the amount of dye transported to the surface.

Nernst's model was further developed by Langmuir (1912) who postulated that near to the interface a static film of liquid exists with different physico-chemical properties from those of the bulk liquid. Langmuir also assumed that the diffusion across the static film is the rate determining process in heterogeneous reactions and that the tangential movement of fluid does not influence the rate of diffusion.

Experimental results, however, appear to contradict the existence of a static layer near solid surfaces. Thus the liquid movements were observed even at about 10^{-5} cm in distance from a surface (Fage and Townsend, 1932). A further weakness of both Nernst's and Langmuir's theories is that neither of them can predict the thickness of the "static layer" on the basis of the properties of fluid and its motion. These theories can only be regarded as semi-empirical.

A more profitable approach for describing mass transfer from a flowing liquid is to solve the differential equation of convective mass transfer (Levich, 1962).

In a moving liquid the mass transport is the sum of two flows: the convection flow,

$$J_{con} = CV \qquad (3.165)$$

and the molecular diffusion flow,

$$J_{diff} = D \, \text{grad} \, C \qquad (3.166)$$

where C is the concentration, D the diffusion coefficient and V the velocity of flow. Since material is conserved

$$J_{conv} = J_{diff} \qquad (3.167)$$

Far from the solid surface, where the velocity of flow is considerable, the diffusional flow can be neglected. However near the surface, where the movement of the fluid slows down, the diffusional mass transport will play an increasingly important role. Thus the liquid can be divided into two regions: the bulk where the solute concentration is uniform, and is essentially independent of the space coordinates (convective flow predominates), and the boundary region where the diffusion flow is important and the concentration varies from point to point.

The thickness of the diffusion layer can be calculated after solving (3.165) to

(3.167) and can be expressed, in a good approximation by

$$\delta \approx \left(\frac{D}{V}\right)^{1/3} \delta_0 \qquad (3.168)$$

where δ_0 is the hydrodynamic boundary layer, see (3.162).

The diffusion flux across the diffusion layer can be expressed as

$$J_{\text{diff}} = \frac{D C_0}{\delta} \qquad (3.169)$$

Thus the quantity δ is essentially the thickness of the Nernst boundary layer. By means of the present theory we can calculate δ from the physical properties of the liquid and of the velocity of flow.

Furthermore, unlike the Nernst boundary layer, the diffusion layer cannot be regarded as a static film, it is only a mathematical concept. The value of δ depends on the geometry of the solid body and on the value of diffusion coefficient. Thus, if several materials diffuse simultaneously, several diffusional layers can be defined.

D   Convective diffusion in turbulent flow

When turbulent flow exists in the vicinity of the surface, it can be subdivided into three regions: (Fig. 3.36)

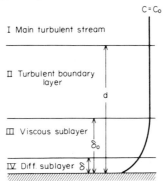

Fig. 3.36   Schematic representation of the various zones at a solid liquid interface when the streams past the surface. (Reproduced with permission from Levich, 1962.)

   i. main turbulent stream where the flow is governed entirely by turbulence

   ii. turbulent boundary layer where there is a transition between the turbulent flow of the main stream and laminar flow in the viscous layer

   iii. viscous sublayer where the flow is laminar but mass transport is governed by molecular diffusion.

The exact expression of the material flux from a turbulent stream depends on the geometry of the solid-liquid interface (for examples see Levich, 1962). Turbulent flow rarely occurs in dyeing processes and therefore the mass transfer from the turbulent stream does not warrant further consideration here.

E   Role of liquid flow in dyeing

Relatively few experimental studies have been carried out on the effect of liquid flow on dyeing. Most of the results obtained suggest that, unless the liquid flow is very slow relative to the substrate, the rate of dyeing is governed by the rate of transport of the dye in the substrate. The geometry of fibres and assemblies of cloth are also much more complex that the idealized cases discussed above and the equations in the previous section can serve only as guidelines.

Rideal (1954) assessed the importance of geometrical factors on the rate of dyeing by comparing the times required for achieving half dye saturation $(t_{1/2})$ of cellophane slabs and of rayon fibres at various bath agitation rates. The effect of stirring was much more important with the fibrous substrate than with the simple cellophane slabs.

Alexander and Hudson (1950) studied the dyeing rates of wool as a function of stirring speed of the dye bath. The dyeing rate as measured by $t_{1/2}$, the half time of complete dye saturation with n, the agitation speed, but approached a limiting value (Fig. 3.37).

At a given stirring speed, the dyeing rates were increasing with the initial dye concentrations (Fig. 3.37). Alexander and Hudson fitted their data to an empirical equation giving the dye uptake as a function of time (Fig. 3.38).

$$\log\left(\frac{C_0}{C_t}\right) = \left(\frac{\beta n}{D}\right)t \qquad (3.170)$$

C_t is the concentration of dye left in the bath after time of dyeing, C_0 the initial dye concentration, D the diffusion coefficient of the dye, n the stirring rate in R.P.M., β a constant and t the time of dyeing.

More recently McGregor and Peters (1965) studied the effect of liquid flow velocity in the rate of dyeing point of view. Assuming that the thickness of the diffusional boundary layer can be expressed by simple inverse function of the flow velocity, V_0,

$$\delta \propto \left(\frac{1}{V_0}\right) \qquad (3.171)$$

they calculated the time of half dyeing of a sheet and a cylinder which is bathed by a streaming dye solution, and obtained good agreement between theory and experiment (Fig. 3.39).

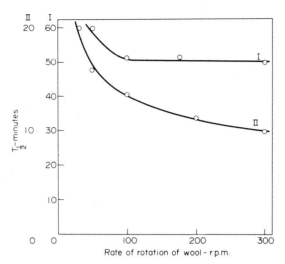

Fig. 3.37 Effect of stirring on $t_{1/2}$ for dyeing of wool with Orange II. Curve I dye concentration 500 mg/l; Curve II dye concentration 125 mg/l. Temperature 25°C. (Reproduced with permission from Alexander and Hudson, 1950.)

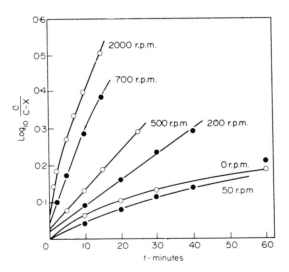

Fig. 3.38. Plot of dye uptake data at various stirring speeds according to equation 3:170. (Reproduced with permission from Alexander and Hudson, 1950.)

XI   DYEING FROM ORGANIC SOLVENTS

In recent years a great deal of interest has been shown in the dyeing of textile fibres with processes which employ organic compounds instead of water or solvents (Milicevic, 1970). Duffy and Olson (1972) compared the kinetics of dyeing of acrylic fibres from aqueous and organic solvents. They concluded

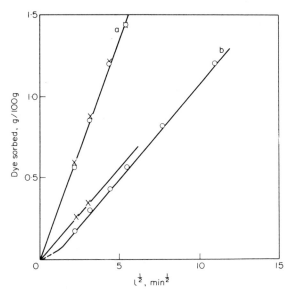

Fig. 3.39.   Dye uptake as a function of dye concentration. Points experimental, curves calculated. For details see McGregor and Peters (1965).

Dye concn
(g/l)

a	1·0	⊙ unstirred
b	0·1	× vigorously stirred

Dye: Chlorazol Sky Blue FF, in presence of 5·0 g/l NaCl Cellulose: Cellophane non-waterproof PT 300 film. (Reproduced with permission.)

that the mechanism of dye transport is different when, instead of water, tetrachloroethylene or m-xylene is used. Whereas in aqueous systems the dyeing kinetics appear to be in accordance with the free volume theory of dyeing, in organic solvents the rate of dye uptake is according to the rigid pore theory. The values of dye diffusion coefficients for given dyes were also found lower by a factor 30 to 60 than those measured in a purely aqueous system.

XII   REFERENCES

Alexander, P. and Hudson, R. F. (1950) *Text. Res. J.*, **20**, 481.

Atherton, E., Downey, D. A. and Peters, R. H. (1955) *Text. Res. J.* **25**, 977.

Atherton, E. and Peters, R. H. (1956) *Text. Res. J.* **26**, 497.

Barrer, R. M. (1941) *Diffusion in and Through Solids*, University Press, Cambridge.

Bell, J. P. (1968) *J. Appl. Polym. Sci.* **12**, 627.

Blacker, J. G. and Patterson, D. (1969) *J. Soc. Dyers and Colourists*, **85**, 598.

Boltzmann, L. (1894) *Annal. Phys. Lpz.*, **53**, 959.

Boulton, H., Delph, A. E., Fothergill, F. and Morton, J. H. (1933), *J. Text. Inst.* **24**, 113.

Braun, G. and Kovacs, A. J. (1965) *Proc. Conf. Phys. Non Crystalline Solids.*, ed. J. A. Prinz, North Holland Publ., Amsterdam.

Breuer, M. M. (1964) *Trans. Far. Soc.*, **60**, 1003.

Breuer, M. M. (1965) *III Congress Internationale de la Recherche Textile Laniere* Vol. 3, p. 91.

Bueche, F. (1956) *J. Chem. Phys.*, **24**, 418.

Bueche, F. (1959) *J. Chem. Phys.*, **30**, 748.

Bueche, F. (1962) *J. Chem. Phys.*, **36**, 2940.

Butcher, B. H. and Cussler, E. L. (1972) *J. Soc. Dyers and Colourists*, **88**, 398.

Chantrey, G. and Rattee, I. D. (1969) *J. Soc. Dyers and Colourists*, **85**, 618.

Cohen, M. H. and Turnbull, D. (1959) *J. Chem. Phys.*, **31**, 1164.

Crank, J. (1956) *The Mathematics of Diffusion*, The Clarendon Press, Oxford.

Crank, J. and Park, G. S. (1968) *Diffusion in Polymers*, Academic Press, London.

Cussler, E. L. and Breuer, M. M. (1972a) *Nature, Physical Science*, **235**, 74.

Cussler, E. L. and Breuer, M. M. (1972b) *A.I.Ch.E. Journal*, **18**, 812.

Davis, C. W. and Shapiro, P. (1964) *Encyl. Polym. Sci. Technology*, Interscience, New York, p. 342.

Daynes, H. (1920) *Proc. Roy. Soc.*, **97A**, 286.

de Groot, S. R. and Mazur, P. (1962) *Thermodynamics of Irreversible Processes*, North Holland Publ. Co., Amsterdam.

Doolittle, A. K. and Doolittle, D. B. (1957) *J. Appl. Phys.*, **28**, 901.

Duffy, E. A. and Olson, E. S. (1972) *J. Appl. Polym. Sci.*, **16**, 1539.

Dumbleton, J. H., Bell, J. P. and Murayama (1968) *J. Appl. Polym. Sci.*, **12**, 2491.

Dumbleton, J. H., and Murayama, T. (1967) *Kolloidzeitschrift*, **220**, 41.

Einstein, A. (1906) *Annal. Phys. Lpz.*, **17**, 549 and **19**, 371.

Fage, A. and Townsend, H. C. H. (1932) *Proc. Roy. Soc.*, **135A**, 656.

Ferry, J. D. (1970) *Viscoelastic Properties of Polymers*, Wiley, New York.

Fitts, D. D. (1962) *Non-Equilibrium Thermodynamics*, McGraw-Hill Book Co., New York.

Flory, P. J. (1953) *Principles of Polymer Chemistry*, Cornell University Press, Ithaca.

Fox, T. G. and Flory, P. J. (1950) *J. Appl. Phys.*, **21**, 581.

Fujita, H. (1961) *Adv. Polymer Sci.*, **3**, 1.

Fujita, H., Kishimoto, A. and Matsumoto, K. (1960) *Trans. Far. Soc.*, **56**, 424.

Garvie, W. M. and Neale, S. M. (1938) *Trans. Far. Soc.*, **34**, 335.

Gilbert, G. A. and Rideal, E. K. (1944) *Proc. Roy. Soc.*, **182A**, 335.

Glansdorff, P. and Prigogine, I., (1971) *Thermodynamic Theory of Structure, Stability and Fluctuations*, Wiley Interscience.

Glasstone, S., Laidler, K. and Eyring, H. (1941) *The Theory of Rate Processes*, McGraw Hill, New York.

Gorkhale, M. K., Peters, L. and Stevens, C. B. (1958) *J. Soc. Dyers and Colourists*, **74**, 236.

Hartley, G. S. and Crank, J. (1949) *Trans. Far. Soc.*, **45**, 801.

Herzog, R. O., Illig, R. and Kudar, H. Z. (1933) *Z. phys. Chem.*, **167**, 329.

Hill, A. V. (1928) *Proc. Roy. Soc.*, **104B**, 65.

Jaeger, J. C. (1949) *An Introduction to Laplace Transformation*, Methuen, London.

Jost, W. (1952) *Diffusion in Solids, Liquids and Gases*, Academic Press, New York.

Karrholm, M. and Lindberg, J. (1956) *Text. Res. J.*, **26**, 528.

Katchalsky, A. and Curran, P. F. (1965) In *Non-Equilibrium Thermodynamics in Biophysics*, Harvard University Press, Cambridge, Mass.

Langmuir, I. (1912) *Phys. Rev.*, **34**, 401.

Levich, V. G. (1962) *Physiochemical Hydrodynamics*, Prentice-Hall, Englewood Cliffs.

Mandelkern, L. (1964) *Crystallization of Polymers*, McGraw-Hill, New York.

McGregor, R. (1966) *J. Soc. Dyers and Colourists*, **82**, 450.

McGregor, R., and Milicevic, B. (1966) *Helv. Chim. Acta*, **49**, 1302.

McGregor, R., Peters, R. H. and Petropolous, J. H. (1962) *Trans. Far. Soc.*, **58**, 771, 1054.

McGregor, R., Peters, R. H. and Petropolous, J. H. (1961) *J. Soc. Dyers and Colourists*, **77**, 704.

McGregor, R., and Peters, R. H. (1965) *J. Soc. Dyers and Colourists*, **81**, 393.

McGregor, R., Peters, R. H. and Ramachandran, C. R. (1968) *J. Soc. Dyers and Colourists*, **84**, 9.

Medley, J. A. (1957) *Trans. Far. Soc.*, **53**, 1380.

Medley, J. A. (1965) As for Breuer (1965) **3**, 117.

Medley, J. A. (1964) *Trans. Far. Soc.*, **60**, 1010.

Meixner, J. (1966) *Symp. Gen. Networks*, Polytech. Inst. Brooklyn, New York, **16**.

Milicevic, B. (1962) *Chimia*, **16**, 29–56.

Milicevic, B. and McGregor, R. (1966) *Helv. Chim. Acta*, **49**, 1319.

Milicevic, B. and McGregor, R. (1966) *Helv. Chim. Acta*, **49**, 2098.

Milicevic, B. (1969) *Text. Res. J.*, **39**, 677.

Milicevic, B. (1970) *Rev. Progress Colouration*, **49**, 1.

Murayama, T., Dumbleton, J. H. and Williams, M. L. (1968) *J. Polym. Sci.*, **6A2**, 787.

Neale, S. M. and Stringfellow, W. A. (1933) *Trans. Far. Soc.*, **29**, 1167.

Neale, S. M. and Stringfellow, W. A. (1933) *Trans. Faraday Soc.*, **29**, 1167.

Nernst, (1904) *Z. Phys. Chem.*, **47**, 55.

Oster, G., Perelson, A. and Katchalsky, A. (1971) *Nature*, **234**, 393.

Perrin, F. (1936) *J. Phys. Radium*, **7**, 1.

Peters, R. H. and Petropolous, J. H. (1959) *Bull. Inst. Text. Fr.*, **84**, 49.

Peters, L. and Speakman, J. B. (1949) *J. Soc. Dyers and Colourists*, **65**, 63.

Peters, L. and Stevens, C. B. (1958) *J. Soc. Dyers and Colourists*, **74**, 183.

Prigogine, I. (1967) *Introduction to Thermodynamics of Irreversible Processes*. Wiley, New York.

Rideal, E. (1954) *Discuss. Faraday Soc.*, **16**, 9.

Rosenbaum, S. (1963a) *J. Appl. Polym. Sci.*, **7**, 1225.

Rosenbaum, S. (1963b) *Text. Res. J.*, **33**, 899.

Rosenbaum, S. (1964) *Text. Res. J.*, **34**, 52, 159, 291.

Rosenbaum, S. (1965a) *J. Appl. Polym. Sci.*, **9**, 2071.

Rosenbaum, S. (1965b) *J. Polym. Sci.*, **3**, 1949.

Rosenbaum, S. and Goodwin, F. L. (1965) *J. Appl. Polymer Sci.*, **9**, 333.

Sprague, B. S. (1967) *J. Polym. Sci.*, **5C**, 159.

Sutherland, W. (1905) *Phil. Mag.*, **9**, 781.

Svedberg, T. and Pedersen (1940) *The Ultracentrifuge*, Clarendon Press, Oxford.

Tyrrell, H. J. V. (1961) *Diffusion and Heat Flow in Liquids*, Butterworth, London.

Turnbull, D. and Cohen, M. H. (1961) *J. Chem. Phys.*, **34**, 120.

Vickerstaff, T. (1954) *The Physical Chemistry of Dyeing*, Oliver and Boyd, London.

Walker, J. A., (1955) *Expl. Cell Res.*, **8**, 568.

White, M. A. (1968) Ph.D. Thesis, University of Leeds.

Williams, M. L., Landel, R. F. and Ferry, J. D. (1955) *J. Amer. Chem. Soc.*, **77**, 3701.

Wood, L. A. and Bekkedahl, N. (1946) *J. Applied Physics.*, **17**, 362.

Wright, M. L. (1954) *Trans. Far. Soc.*, **50**, 89.

Chapter 4

Dyes in solution

It is clear from a consideration of molecular bonding forces and adsorption that dye molecules are prone to interact not only with adsorbing substrates such as materials to be dyed or sections to be stained, but also with one another and with other solutes which may be present. A dye molecule capable of participating in a hydrophobic interaction with a substrate surface must be capable of participating in an analogous dye-dye interaction. Thus the physical interactions of dye molecules, particularly in aqueous solution, are of importance both at the theoretical level because of thermodynamic considerations and at the practical level where such interactions may be either advantageous or disadvantageous. The important interactions at the theoretical level are those which lead to thermodynamically non-ideal behaviour while at the practical level grosser effects such as aggregation and association are important. The two may be considered separately for convenience.

I THE NON-IDEALITY OF DYE SOLUTIONS

Raoult was the first to note that within the limits of experimental accuracy the partial pressure of a solvent containing a non-volatile solute was proportional to the mol fraction (x) of the solvent,

$$p = p_o x \qquad (4.1)$$

Regarding the vapour pressure of the solvent as being proportional to its concentration at the liquid/vapour interface, this is explained by treating the solute molecules as non-volatile equivalents of the solvent molecules diluting the surface concentration strictly proportionally to their mole fraction. This will be true if the solute-solvent interaction energy is the same as the solvent–solvent interaction energy and the molecular volumes of the solvent and solute are the same. A number of dilute solutions obey Raoult's Law which expresses itself in a whole series of relationships between the effects of solutes and their concentration which describe elevation of boiling point depression of freezing point, osmotic pressure etc. Under most circumstances solutions do not behave in the ideal manner required by Raoult's Law and these solutions are described as being to a greater or lesser extent non-ideal.

In thermodynamic terms the chemical potential of a component i in an ideal solution is given by

$$\mu_i = \mu_i^0 + RT \ln x_i \qquad (4.2)$$

in which μ_i^0 is the standard chemical potential dependent upon temperature and pressure and equal to μ_i when $x_i = 1$, i.e. when the component i is in the pure state. Certain specified non-ideal situations have been considered. If the molar volumes of solvent and solute are different, Hildebrand and Scatchard have suggested that

$$\mu_i = \mu_i^0 + RT \ln x_i + (1 - \phi_i)^2 \, \Delta U^0 \qquad (4.3)$$

in which ϕ_i is the volume fraction of i and ΔU^0 is the exchange energy of interaction of the components. On the other hand, when i is an electrolyte the classical Debye–Huckel–Onsager relationship for small ions in dilute solution gives

$$\mu_i = \mu_i^0 + RT \ln x_i - Bc_i^{1/2} \qquad (4.4)$$

in which B is a positive constant depending on the ionic charge and dielectric constant of the solution c_i is the ionic strength. Clearly the situation becomes very complex with large ions at high concentration etc. The situation becomes much simpler when the purely phenomenological approach of G. N. Lewis is employed. This utilizes a correction term f which has a value such that

$+RT \ln f$ provides the appropriate correction value to the thermodynamic relationship. f is termed the activity coefficient so that

$$\mu_i = \mu_i^0 + RT \ln x_i + RT \ln f_i = \mu_i^0 + RT \ln x_i f_i \qquad (4.5)$$

The product $x_i f_i$ was termed by Lewis the activity of i. This is clearly the mol fraction of an ideal solute giving the same chemical potential as that given by i when its mol fraction is x_i. It can also be regarded as the thermodynamic or behavioural mol fraction of i. The value of f_i may be determined if the extent of the non-ideal behaviour of the solute i can be measured. It is not always necessary, of course, to employ mol fractions to describe the amount of solute present since in dilute solution that function is directly proportional to concentration. Thus x_i in equation (4.5) may be replaced by the concentration c_i. This has the effect of changing μ_i^0 to the value of μ_i when $c_i f_i = 1$. There is an extensive literature relating to the measurement of activity coefficients. This is considered in the present context particularly from the point of view of the special problems associated with dye solutions.

A   The determination of activity coefficients of dyes in solution

The most direct method of determining the activity of a solute is through vapour pressure measurement of solutions. This provides an activity coefficient for the solvent so that the corresponding value for the solute may be calculated using the Gibbs–Duhem equation.

However the zone of interest in dye solutions is in the range 10^{-3}–10^{-6} molar and at these concentrations the value of p/p_0 in Raoult's equation is between 0·99998 and 0·9999998 representing pressure differences of $1·37 \times 10^{-2} - 1·37 \times 10^{-5}$ mm. Regular manometry is clearly of little use in these circumstances since activity changes to be measured involve detection of pressure differences much smaller than these values. Special methods of amplification must be employed. These fall into three categories.

1. Static methods

This is a form of direct manometry which employs a wide bore manometer so that the volume difference between the two sides is very much greater than the difference in height. The volume difference determined in terms of manometric height in a fine capillary reflects an amplification which relates to the cross sectional area of the two manometer fibres. This is illustrated in Fig. 4.1.

This technique is due to Puddington (Sirianni and Puddington, 1956) and can give amplification of 2500–4000 with high accuracy. Thus pressure differences of the order of 5×10^{-2} mm require to be measured in the most difficult cases only. The method works well with non-aqueous solvents but difficulties using aqueous solutions are reported. Aston (1970) has produced activity data for azobenzene using this technique and has shown the temperature dependence of activity coefficients leading to a minimum value at temperatures near 38°C.

Another approach is to use radioactive labelling of the solvent where appropriate. By prolonged counting the amount of solvent in the vapour phase can be determined with high precision. This method has been used for the determination of vapour pressures of some compounds (Rosen and

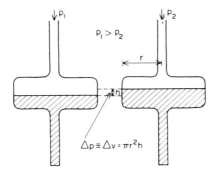

Fig. 4.1. A schematic version of the vapour pressure amplification system of Puddington.[1]

Davies, 1953). It clearly offers difficulties in the case of the vapour pressure of aqueous solutions of non-volatile dyes and also counting disintegrations over long periods introduces many sources of error.

2. Dynamic methods

These are based on the fact that if a gas is passed through a liquid it will cause a loss in volume at a rate proportional to the vapour pressure. The evaporated liquid may be collected and the rate of removal determined. Care must be taken to ensure the gas bubbles become saturated and that no mechanical losses occur. The method has been used with electrolyte solutions and examined in a preliminary way with dye solutions (Bechthold and Newton, 1940). The vapour pressure of volatile dye solids has also been studied by Friedrich and Stammbach (1964, 1968). By prolonging the duration of the experiment differences in rates of evaporation can be amplified but at the same time possible errors due to non-uniformity of gas flow rates, temperature variations etc., are also increased.

3. Isopiestic methods

These are based on some equilibrium or steady state condition relating to two systems in contact, the properties of one being known. For example, if two droplets of solvent and solution are formed near to one another, solvent will evaporate more rapidly from the droplet of pure solvent due to its higher vapour pressure and it will tend to condense on the droplet of solution. The processes of evaporation and condensation will lead to temperature differences which may be expressed electrically using thermistors and amplified. This approach is incorporated in the Vapour Pressure Osmometer (Burge, 1963). Water is not a very suitable solvent for this method due to the high latent heat of vaporization but studies have been carried out by Milicevic and Eigenmann (1964).

Since the transfer of solvent between the two droplets is a rate process, time may be used also as the "amplifier". If a slow rate of change is integrated over a long time it can be determined with accuracy. This approach has been employed also in isopiestic vapour pressure studies with dye solutions by Chadwick and Neale (1958). The principle of the method are illustrated in Fig. 4.2.

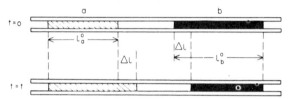

Fig. 4.2.   Capillary isopiestic vapour pressure determination.

The diagram shows the position at the start $(t = 0)$ with two solutions, a and b, of known molality and after a period of time $(t = t)$ when equilibrium has been reached and the vapour pressures are equal.

If solutions a and b are of known initial concentrations, c_a^0 and c_b^0 respectively, and the activity coefficient of b is known for a range of concentrations then at equilibrium,

$$\text{the activity of } a = c_a^t f_a^t = \frac{l_a^0}{l_a^0 + \Delta l} . c_a^0 f_a^0 \tag{4.6}$$

$$\text{and the activity of } b = c_b^t f_b^t = \frac{l_b^0}{l_b^0 - \Delta l} . c_b^0 f_b^0 \tag{4.7}$$

and since at equilibrium when the two vapour pressures are equal the activities must be equal,

$$f_a^t = \frac{l_b^0}{l_a^0} . \frac{l_a^0 + \Delta l}{l_b^0 - \Delta l} . \frac{c_b^0}{c_a^0} . f_b^t \tag{4.8}$$

The big problem with this method is bubble formation due to air dissolved in the solvent when water is employed. The use of other solvents presents no such difficulty although the method generally requires manipulative skill. Chadwick and Neale (1958) operated on a large scale using two interconnected tubes (Fig. 4.3).

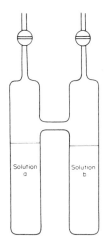

Fig. 4.3. Schematic representation of Chadwick and Neale's apparatus.

The principles of the larger apparatus are the same. Air bubbles present much less difficulty but due to the greater quantity of solvent to be transferred, experimentation time and hence temperature stabilization problems are increased. A few days will suffice to equilibrate solutions in acetone or chloroform but weeks or months are needed with aqueous solutions.

4. *Other methods for the study of dye activity*

One of the normal methods in studies of this kind is the determination of the elevation of boiling point. Due to the high dilution of dye solutions the same problem arises as with the direct measurement of vapour pressure. The boiling point elevation due to a 10^{-4} molar concentration of an ideal solute in water at 25°C is 3.2×10^{-5}°C. In order to measure activity coefficients with any reasonable accuracy temperature would require to be measured to within $\pm 1 \times 10^{-7}$°C which is not possible at the present time.

The most frequent practical activity determination is the measurement of pH. When an electrode adsorbs ions reversibly a characteristic potential ψ is set up whereby,

$$\psi = \psi_0 - \frac{2RT}{F} \ln a \qquad (4.9)$$

in which a is the activity of the ion. A number of specific ion electrodes are available for simple ions such as NH_4^+, I^-, Br^-, Cl^- and S^- as well as sodium ions and protons. The reversibility in the context of equation (4.9) is electrical rather than chemical and due to their powerful physical adsorption properties the development of a reversible electrode for dye ions has not proved possible. Indeed the physical adsorption of dye ions can interfere with regular determinations so that pH measurement in free dye acid solution may often be found to be inaccurate unless the glass electrode is pre-conditioned in dye solution.

II   THE AGGREGATION OF DYES IN SOLUTION

If the study of the thermodynamic activity of dyes in solution presents very great difficulty, the phenomenon of dye aggregation is more amenable to experimental study in terms of the magnitude of the effects to be observed. However it will be seen that considerable difficulties of interpretation remain.

The formation of an aggregate by dye molecules implies the direct operation of molecular binding forces and the development of a series of equilibria in solution

$$2D \; \overset{K_1}{\rightleftharpoons} \; D_2$$

$$D + D_2 \; \overset{K_2}{\rightleftharpoons} \; D_3$$

$$D + D_3 \; \overset{K_3}{\rightleftharpoons} \; D_4$$

$$\text{or } 2D_2 \; \overset{K_4}{\rightleftharpoons} \; D_4 \text{ etc.} \tag{4.10}$$

when the dye molecules ionize in solution, the situation is rendered more complex by the different ways in which the co-ions can be associated with the aggregates. The phenomenon can be studied in a number of ways.

A   Methods based on aggregate size

Since the formation of a dimer or higher aggregate involves a considerable increase in molecular weight and size diffusion controlled properties provide a basis for the study of aggregation.

In the direct study of diffusion properties the rate of flux of dye molecules down a concentration gradient is determined by appropriate means and the diffusion coefficient calculated by applying the data to Fick's Law

$$ds/dt = -D . \nabla c \tag{4.11}$$

the value of D may be related to molecular size using the Stokes–Einstein relation (assuming that the molecules are spherical),

$$D = \frac{RT}{6\pi\eta N} \cdot \frac{1}{r} \qquad (4.12)$$

in which η is the viscosity, N is Avogadro's number and r is the radius of the diffusing species. With uncharged molecules an estimate of the degree of association can be made from the value r.

The methods which may be used are based upon bringing two solutions, one containing the dye, into contact either directly by putting the dye solution into a capillary tube and immersing the tube in a dye free solution or indirectly by separating the two solutions by a glass sinter. In the latter case static and dynamic methods are available.

In the static method, represented in Fig. 4.4, the rate of transport of dye across the sinter is measured spectrophotometrically either in a time lag or steady state diffusion experiment. The dynamic method is a convenient means of producing accurate results rapidly. It is shown schematically in Fig. 4.5.

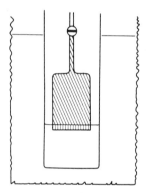

Fig. 4.4.   A schematic representation of a simple diffusion cell for the study of dye diffusion in solution.

In the dynamic method dye-free solvent flows at a known rate through the compartment into which dye is diffusing. The diffusion rate across the sinter is given by equation 4.11. If the dilution rate of the upper compartment is suitably adjusted then a steady state can be attained when the concentration rate (or ds/dt in equation 4.11) equals the dilution rate, i.e. if C_s is the steady state concentration, the dilution rate is v l.min^{-1} and the volume of the upper compartment is V

then
$$\frac{ds}{dt} = \frac{C_s v}{V} \qquad (4.13)$$

The value of C_s, v, V and the concentration in the bottom compartment are all readily determined. The concentration gradient must be calculated using a calibration experiment because the true diffusional area and thickness of sinter cannot be measured directly. If a substance of known diffusion coefficient in the solvent is used these parameters are readily calculated.

The value of the dynamic method rests in the speed of achieving the steady state and the rapidity of repeat measurements. The latter are made by drawing some of the solution from the upper compartment into the lower. This leads to a dilution which disturbs the steady state condition.

The difficulty with diffusion methods is that most dyes exist as ions in aqueous solution. In the diffusion process small counter-ions, Na$^+$ or H$^+$

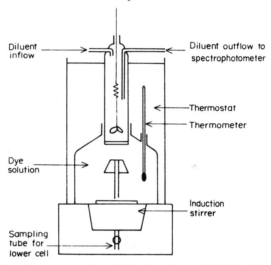

Fig. 4.5. A schematic representation of the dynamic method of studying dye diffusion in solution.

move much more rapidly than the dye ions across the interface. The resultant charge separation leads to a diffusion potential which slows down counter-ion motion but accelerates that of the dye ion. The observed coefficient is then an average value given by the Nernst–Haskell equation

$$D = \frac{RT}{F}\left\{\frac{1}{n_d}+\frac{1}{n_i}\right\}\frac{V_d V_i}{V_d+V_i}\times 10^{-7} \tag{4.14}$$

in which V_d and V_i are the mobilities of the dye and counterions while n_d and n_i are the corresponding charges. The mobility of the dye ion is readily obtained from conductivity data since

$$\Lambda = F(V_d+V_i) \tag{4.15}$$

and the value of V_i is obtainable from the literature on the basis of the study of simple electrolytes. The difficulty arises from the uncertainty about the value of n_d. In the formation of an aggregate counterions may be included so that the formal charge on the aggregate is different from the actual value. This was shown by Robinson and Moilliet (1934) who determined the transport number of dye ions (t_d) as well as conductivity. This enables V_d to be calculated directly.

$$t_d = \frac{F V_d}{\Lambda} = \frac{V_d}{V_d + V_i} \qquad (4.16)$$

From the values of V_d the apparent value of V_i may be calculated and compared with determined using simple electrolyte solutions. In the three cases studied Robinson and Moilliet found V_i to be unexpectedly low and that the discrepancy increased with dye concentration. Thus the dye aggregates behaved as if sodium ions were included.

The problem of the diffusion potential can be avoided if the diffusion behaviour is studied in an electrolyte solution sufficiently strong to make the charge separation effect negligible. Robinson (1935) showed marked aggregation of Benzo purpurine 4B in this way. A concentration of salt in excess of 0·03N was necessary to negate the diffusion potential. The aggregation number was calculated from the particle radius obtained from the Stokes–Einstein equation. Valkó (1937) carried out an extensive study and a typical group of his results are shown in Table 4.1. The expected increase in aggregation due to the addition of salt can be seen. Similar experiments were carried out by Lenher and Smith (1935) who noticed that in many cases the degree of aggregation of dyes increased with time. This effect has been confirmed more recently in another case (Coates, 1969). The effect of temperature has been examined by a number of investigators with the general conclusion that high temperatures promoted disaggregation. Temperature studies are difficult using a sintered glass diffusion "membrane" due to dimensional changes in the glass. The errors are not large however and satisfactory measurements can be made.

An alternative approach to the use of higher electrolyte concentrations to offset diffusion potentials is the use of radioactive labelling of dye molecules so as to study the diffusion of dye under conditions of zero net flux. This technique has not been applied to solution studies although it has been used to study diffusion in polymers.

Another diffusion based technique uses polarography (Hillson and McKay, 1965) when two electrodes are placed in an aqueous solution and a potential difference applied, the anode tends to accept electrons from the solute leading to oxidation, while at the cathode reduction occurs. If the cathode is a dropping mercury electrode the products of the reduction are prevented from accumulating. Reduction (and oxidation) at the electrode depends upon the

TABLE 4.1. Diffusion coefficients, particle sizes and aggregation numbers for direct dyes in water at 25°C (Valkó, 1937).

Dyestuff	Dye concentration (g/l)	Salt concentration (molar)	Diffusion coefficient ($m^2 \, sec^{-1}$)	Calculated particle radius (nm)	Aggregation number
Benzo purpurine 4B	0·5	0·01	$2·64 \times 10^{-10}$	1·01	6
	0·10	0·01	$2·49 \times 10^{-10}$		
	0·05	0·02	$2·18 \times 10^{-10}$	1·13	8
	0·10	0·02	$2·16 \times 10^{-10}$		
	0·3	0·03	$1·12 \times 10^{-10}$	2·16	50
	0·2	0·05	$6·02 \times 10^{-11}$	4·04	400
Congo Red	0·5	0·20	$1·23 \times 10^{-10}$	1·98	45
Chicago Blue 6B	0·2	0·50	$1·18 \times 10^{-10}$	2·06	37

applied potential so that as the potential is changed, a sufficiently negative potential at the cathode leads to a ready acceptance of electrons by solute molecules and a large increase in current with a small change in potential. When the negativity of the potential is sufficient to bring about reduction then this process leads to the depletion of the solution around the dropping mercury electrode so that the rate of reduction and hence the current depends upon the diffusion of the solute towards the electrode. The relationship between the factors operating is given by the Ilkovic equation

$$i_d = K \, n \, D^{1/2} \, c \, m^{2/3} \, t^{1/6} \tag{4.17}$$

in which i_d is the diffusion current, K is a system constant, c is the concentration of the solute, m is the weight of the mercury drop and t is the drop formation time. More rigorous forms of the Ilkovic equation can be developed. Hillson and McKay (1965) showed that a general phenomenological relationship between diffusion coeffients and molecular weights could be used to calculate apparent molecular weights and aggregation numbers from polarographic diffusion data. The technique is not entirely free from experimental problems. Adsorption effects in the polarographic cell can be seen from time to time. However the association behaviour of a number of dyes has been examined. In Table 4.2 below typical data obtained by Coates (1969) with the dye 1-(2'-hydroxy-5'-sulpho-phenylazo)-2-hydroxynaphthalene (C.I. 15670) at 25°C.

TABLE 4.2. Typical aggregation data from polarographic studies (Coates, 1969).

Concentration (molar)	Diffusion coefficient ($m^2 \, sec^{-1}$)	Apparant mol. wt.	Aggregation number
$4·55 \times 10^{-5}$	$4·47 \times 10^{-10}$	347	1·00
$9·10 \times 10^{-5}$	$4·46 \times 10^{-10}$	347	1·00
$1·52 \times 10^{-4}$	$4·37 \times 10^{-10}$	363	1·04
$6·82 \times 10^{-4}$	$4·00 \times 10^{-10}$	457	1·31
$1·36 \times 10^{-3}$	$3·83 \times 10^{-10}$	525	1·51
$2·73 \times 10^{-3}$	$3·47 \times 10^{-10}$	661	1·90

The diffusion coefficients are proportional to the square of the diffusion current and the apparent molecular weight values are very sensitive to the value of the diffusion coefficient as can be seen in Table 4.2. Consequently the value of the aggregation number is sensitive to errors in measuring i_d.

B   Methods based on spectrophotometry

In many cases due to the participation of the delocalized electrons in dye molecules in the formation of dye–dye complexes there is an observable change in the electronic absorption spectrum. It is not possible to determine from such changes the state of aggregation under equilibrium or steady state conditions although special spectrophotometric techniques have been used to study the rate processes in a particular case and hence the aggregation number.

Ultraviolet and visible absorption spectra provide a very convenient qualitative method of studying aggregation in appropriate cases and a considerable number of examples is to be found in the literature. A typical case is shown in Fig. 4.6.

An interesting example of the use of visible spectrophotometry is provided by Coates (1969) who has shown how the state of aggregation of the dye 1-(2'-hydroxy-4'-sulpho-5'-nitro naphthylazo)-2-hydroxy naphthalene is very dependent upon time.

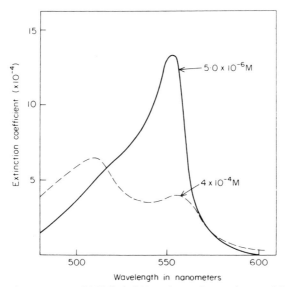

Fig. 4.6.   Absorption spectrum of 3·3'-diethylthiacarbocyanine p-toluene sulphonate in water (West and Pearce, 1965).

The interpretation of electronic absorption spectra in the ultraviolet and visible region is possible only if specific assumptions are made about the state of aggregation.

If the equilibrium is assumed to be a dimerization the situation will be described by the stoichiometric equation

$$2D_1 \overset{Kd}{\rightleftharpoons} D_2$$

where D_1 and D_2 are the monomeric and dimeric forms. If the observed optical density of dye solutions corrected to a standard path length is

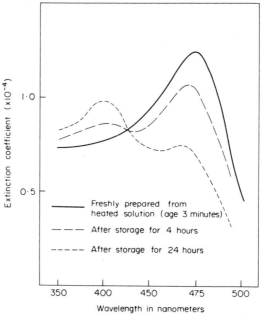

Fig. 4.7. The effect of standing time on the aggregation of a strongly aggregating dye (Coates, 1969).

divided by the nominal molar dye concentration assuming all the dye is in the monomeric form, two limiting values will be obtained. At high dilution the value ε_1 will equal the extinction coefficient of the monomer. The value at high concentration will be half of the extinction value of the dimer ε_2, which may be readily calculated as a consequence. The apparent extinction coefficient for the intermediate situations may be denoted by ε^A. Thus

$$\varepsilon^A_{max} = \varepsilon_1 \; ; \; \varepsilon^A_{min} = \varepsilon_2/2 \tag{4.18}$$

If $[\bar{D}]$ is the total dye concentration expressed as monomer,

$$[\bar{D}] = [\bar{D}_1] + 2[D_2] \tag{4.19}$$

Also $$\varepsilon^A[\bar{D}]=\varepsilon_1[D_1]+\varepsilon_2[D_2] \tag{4.20}$$

Combining equations 4.19 and 4.20 gives

$$\frac{[\bar{D}_1]}{[\bar{D}]}=\frac{2\varepsilon^A-\varepsilon_2}{2\varepsilon_1-\varepsilon_2} \tag{4.21}$$

Thus $[D_1]$ may be calculated and hence $[D_2]$ from equation 4.19 and the dimerization constant from

$$[D_2]/[D_1]^2 = K_d \tag{4.22}$$

The value of $-RT \ln K_d$ gives ΔG_d^0, the partial molar free energy of dimerization. Typical values are given in Table 4.3.

TABLE 4.3.   Typical values of the partial molar free energy of the dimerization of dyes in aqueous solution.

Dye	$-\Delta G_d^0$ (k.cal/mole).
C.I. Acid Orange 7 (Orange II)	4.28 (25°C) (Milicevic and Eigenmann, 1964)
C.I. Basic Orange 14 (Acridine Orange)	5.50 (25°C) (Lamm and Neville, 1965)
3,3'-diethylthiacarbocyanine p-toluene sulphonate	6·03 (22°C) (West and Pearce, 1965)

It can be seen from Table 4.3 that dimerization energies can be quite significant and comparable with the bonding energies with substrates. The effect of dimerization on the adsorption isotherm can be calculated in a particular case.

If it is assumed that both monomer (D_1) and dimer (D_2) are adsorbed in accordance with a partition isotherm with coefficients K_1 and K_2 respectively, and the dimerization is governed by a dimerization constant Kd then the following relationships will apply. Using the subscripts s and f to indicate whether the dye is in the solution or on the fibre respectively,

$$(D_1)_f = K_1(D_1)_s \tag{4.23}$$

$$(D_2)_f = K_2(D_2)_s \tag{4.24}$$

$$(D_2)_s = K_d(D_1)_s^2 \tag{4.25}$$

Also the amount of dye on the fibre $(\bar{D}_f)$ or in solution $(\bar{D}_s)$ will be given by

$$(\bar{D}_f) = (D_1)_f + 2(D_2)_f \tag{4.26}$$

$$(\bar{D}_s) = (D_1)_s + 2(D_2)_s \tag{4.27}$$

Combining equations (4.25) and (4.27) gives

$$(D_1)_s = \{[1+8Kd(\bar{D}_s)]^{1/2}-1\}/4Kd \tag{4.28}$$

and

$$(D_2)_s = \{[1+4Kd(\bar{D}_s)]\pm[1+8Kd(\bar{D}_s)]^{1/2}\}/8Kd \tag{4.29}$$

Hence

$$(\bar{D}_f) = \frac{K_1}{4Kd}\{[1+8Kd(\bar{D}_s)]^{1/2}-1\}$$

$$+\frac{K_2}{4Kd}\{[1+4Kd(\bar{D}_s)][1+8Kd(\bar{D}_s)]\}^{1/2} \qquad (4.30)$$

Thus equation (4.30) is the composite isotherm equation which no longer reflects the partition effect. In Fig. 4.8 the observed isotherms for the above situation where K_1, K_2 and K_d have the values 10^6, 10^4 and 10^5 respectively.

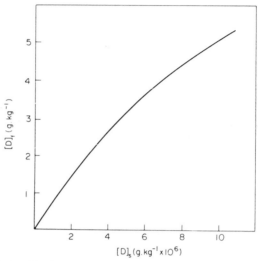

Fig. 4.8. The calculated isotherm for the adsorption of a dimerizing dye where monomer and dimer obey a partition isotherm equation (Equation 4.30).

The dangers of attempting to interpret an adsorption isotherm mechanistically are re-emphasized by this effect.

A sophisticated development of spectrophotometry, the temperature–jump–relaxation technique may be used to provide less ambiguous information. In this technique a very rapid heating pulse is applied to a system in a state of temperature sensitive equilibrium. The consequent spectral changes are followed using special optical systems designed to detect very small changes. One major advantage of the method is that the data may be linked to an iterative computer program to calculate the aggregation and disaggregation velocity constants. The nature of the reaction kinetics provides information additionally about the aggregation number. In this way Hague et al. (1971) were able to show that at 20°C, C.I. Acid Red 66 exists in the form of the dimer to the extent of about 66% even at a concentration as low as 3×10^{-5} molar in water. It is noteworthy that normal spectrophotometric studies using C.I. Acid Red 66 fail completely to show any abnormal behaviour at such concentrations.

Nuclear magnetic resonance spectroscopy has been used to study aggregation by Blears and Danyluk (1966). The formation of molecular complexes may in appropriate cases provide shielding effects leading to field shifts. The form of the aggregation may be deduced from the results. This technique has been little used.

Emission spectra have been used to demonstrate and study aggregation effects. Normally aggregation will lead to a shift in the emission wavelength. The behaviour of different kinds of aggregate is distinguishable by this technique. Where the aggregating molecules adopt a stacked or sandwich configuration a shift of λmax to shorter wavelengths is expected but fluorescence should be at least partially quenched. This behaviour is shown by Acridine Orange (Coates, 1969) (Fig. 4.9). When the aggregates are formed by an "end-on" interaction aggregation should lead to a shift of λmax to longer wavelengths and the fluorescence should remain relatively strong.

Light scattering has also been used in the study of aggregation (Alexander and Stacey, 1952) but there are indications that with dyes giving rise to aggregates large enough to give accurate light scattering data, the solutions are not stable, i.e. aggregation is time dependent.

III   THE SOLUBILITY OF DYES

The question of dyestuff solubility is a practical one and in certain cases one of theoretical importance. In the case of research studies dyes are used in solution and it is activity determination which is the problem. In practice this is frequently not the case and solubility is an important factor. The disperse dyes, which are applied from a suspension, i.e. a saturated solution, have solubility characteristics of both practical and theoretical importance, since the upper limit of concentration in the external phase determines the upper limit of fibre saturation. In such cases both rate of dissolution and solubility are of importance and these are related in turn to the particle size of the dye crystals and the crystalline form.

The solubility of ordinarily soluble dyes is clearly of practical importance in manufacture since at some stage the dye must be precipitated for separation. The more efficiently this can be done and the more readily the reaction product can be filtered off, the better the process. Again both factors will depend on particle size and crystal habit.

Surprisingly very little attention has been paid to the solid state of dye powders despite its evident importance except in the case of vat dyes which are often applied as dispersions to fabrics. It is now being realized that this is a serious omission and investigations are the subject of work by Biedermann in

Switzerland and by F. Jones. Most practical experience of the effect of the dye solid state is related to disperse dyes and the dyeing of man-made fibres. The topic will be considered in greater detail under that heading.

The solubility of dyes has not been so neglected although it has been long recognized that this is one of the more difficult dye properties to measure. The problem is that dyes tend to be surface active and to give colloidal solutions in

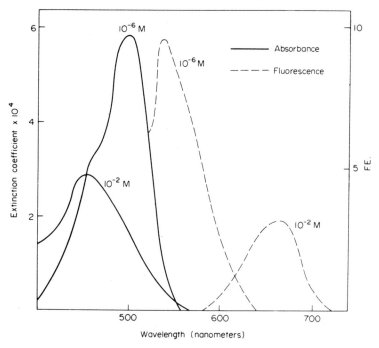

Fig. 4.9.   Absorption and fluorescence spectra of Acridine Orange (Coates, 1969).

high concentrations which resemble true solutions and in some cases they give gels. Another feature is that strong dye solutions are often capable of dispersing in water impurities which are not normally water soluble. Thus it is not possible simply to allow a saturated solution to separate into solution and solid phase, decant and filter. Other methods are not particularly more reliable and the solubility of dyes is normally expressed in terms of some arbitrary test. The sparingly soluble disperse dyes do not present the same problems because at saturation the solutions are very dilute (5–20 ppm). However in this solubility range the value of the saturation solubility is very dependent upon crystal size and habit and both of these are likely to change in the course of the solubility measurement.

IV THE INTERACTION OF DYES WITH OTHER SOLUTES

The interaction of dyes with model substrates in solution is considered elsewhere in relation to dye–carbohydrate dye–protein etc. interactions. The other solutes of interest in the present context fall into the general technical class of textile auxiliaries or dyeing assistants.

The process of aggregation previously discussed will be affected clearly by the presence of other solutes which will lessen dipolar or hydrophobic interactions leading to molecular association. For example disaggregation in the presence of urea, pyridine or dimethylformamide which are all noted hydrophobic interaction inhibitors is a commonplace. The mode of their action in the case of dye aggregation appears to be the formation of an adduct or complex which is soluble and does not aggregate. In the case of dyes dissolved in strong urea solutions, the adduct very slowly precipitates and can be isolated (Gilchrist, 1971) when polyethanoxy compounds of certain types are present. The mechanism is not fully understood in this case but the effect undoubtedly occurs. The use of urea to assist dye solubility is fairly widespread particularly in padding applications and in printing. Observation of the way in which urea is best use, i.e. by mixing with dry dye powder followed by pasting with water, suggests that the formation of the dye–urea adduct is helpful. Urea may be more effective in preventing dye–dye aggregation than in breaking up aggregates once formed.

The interaction of dyes with surfactants present for the purposes of wetting, dispersing, scouring etc. has aroused interest for a considerable period and merits special consideration.

A Dye-surfactant interactions

Dye-surfactant interactions were observed in the 1930's by Valkó (1945) and others. Their study arose from the use of surfactants as dyebath levelling assistants particularly in the dyeing of wool. Not all surfactants were found to show marked interactions with dyes since the ionic agents, e.g. the sulphated fatty alcohols, were not effective. The non-ionic agents on the other hand formed strong complexes with many acid dyes (Schwenn, 1933). The best known practical example was the polyethanoxy compound, Palatine Fast Salt O, which was used to assist the level dyeing on wool of 1:1 chromium complex dyes. Valkó showed by a study of diffusion in solution that a complex was formed in the presence of the agent. The cause of the complexing was not clear however. The aggregation behaviour of one of the 1:1 chromium complex dyes, Palatine Fast Blue GGN, was examined by Valkó (1949) and it was found to *increase* with temperature. This behaviour is also shown by polyethanoxy surfactants and it was later shown by Rattee (1953) that a dye-surfactant complex was formed on warming. The proposed mechanism suggested that on warming the hydrated surfactant molecule "lost water" leading to the formation of an insoluble aggregate; the dye exhibiting a similar tendency entered into complex formation by a process best described as co-operative dehydration. The very close similarity between this concept and that of hydrophobic bonding developed a little later is very clear.

The ability of dyes to form complexes varies widely as indeed does that of surfactants. Craven and Datyner (1967) have observed that the ability to form a complex increases with the hydrophobicity of the dye thus confirming the earlier data. Complexing was found to be inhibited by chain branching in the alkyl chain in the surfactant (Datyner and Delaney, 1971) while it was increased by lengthening the ethylene oxide chain in the molecule. Craven and Datyner (1967) also observed that complex formation was reduced by increasing the temperature with the surfactants studied. Since the effect of temperature increase is to enhance the complex formation in earlier work the effect of temperature can operate either way.

Early studies by Wurzschmitt (1950) led to the proposition that the interaction between the dye and the surfactant occurred between the sulphonic group in the acid dye anion and positively charged oxonium groups in the ethylene oxide chain. However it was later shown by Craven and Datyner (1964) that complexing occurred even when no charged oxonium groups were present. Craven and Datyner (1967) concluded that hydrophobic interactions were involved on the basis of surface tension and cloud point studies thus agreeing with the earlier postulate (Rattee 1953). There is however an apparent contradiction between the hydrophobic interaction mechanism and the increased complex formation effect with longer (hydrophobic) ethanoxy side chains in the surfactant. Craven and Datyner argue that this contradiction is removed if a molecular interaction is proposed between dye and surfactant molecules rather than an interaction between dye molecules and surfactant micelles. The equilibria present are then

surfactant molecules $\rightleftharpoons$ surfactant micelles.
surfactant molecules and dye $\rightleftharpoons$ surfactant-dye complex

The optimum concentration of surfactant molecules (critical micelle concentration) is known to increase with the length of the polyethanoxy side chain so that the first equilibrium is forced to the left and the second to the right. The concept of a molecular interaction is supported by spectrophotometric studies by Nemoto and Imai (1959) as well as surface tension studies. It is also supported by studies of the effect of inorganic salts on complex formation (Rattee, 1953) when it was found that this increased in accord with the lyotropic series and the tendency to re-disperse coacervates. The relevant data are shown in Table 4.3 which relates to the complex formation between C.I.

Acid Green 35 and an octyl-phenol-ethylene oxide condensate (Lissapol N). The electrolyte concentration in each case was 0·1 molar.

TABLE 4.3.   The effect of added electrolyte on complex formation between C.I. Acid Green 35 and Lissapol N (Rattee, 1953).

Added electrolyte	Relative complex formation (no addition ≡ 100)
Sodium borate	11
Magnesium sulphate	81
No addition	100
Sodium thiosulphate	105
Sodium nitrate	116
Sodium acetate	178
Lithium thiocyanate	203
Ammonium thiocyanate	214

The effect of the thiocyanate salts is clearly consistent with their known effect on the solubility of lyophilic substances, i.e. to increase the degree of molecular dispersion and supress micelle formation. Magnesium and sulphate ions can be seen to combine their expected opposite effects and supress complex formation.

Complexing with surfactants has been put to useful effect not only in relation to normal dyeing processes. It has been used to enhance the fastness of dyeings when they are washed in detergents. The Carbolan dyes were introduced by I.C.I. in the 1930's and were notable for the lack of backstaining they produced on white wool when dyed materials were treated in soap solutions. The dyes are regular sulphonated acid dyes which bear a dodecyl- or two butyl-side chains and this group has been shown to confer the property of entering soap micelles to give a dye-soap complex of virtually no affinity for wool. The more recent uses of dye-surfactant interactions have been reviewed by Valkó (1972).

B   Disaggregating agents

In view of the tendency of dyes to aggregate interest has always existed in compounds assisting solution and disaggregation. Strongly dipolar solutes such as urea or dimethylformamide have been extensively used particularly in practical application to assist dissolution of dyes. In recent years the role, particularly, of urea in this respect has been attributed to its water structuring effect almost as often as to its water-destructuring effect! It has been pointed out fairly forcibly by Holtzer and Emerson (1969) that aqueous solutions are frequently too complex for naive applications of such concepts. This is almost certainly true when such complex solutes are involved.

Any practical dyer whose problem is to dissolve large quantities of dyes in relatively small amounts of water knows that urea is of great benefit in assisting dissolution of the dye. This practical day to day observation is supported by numerous more conventional research observations due to Kissa (1969), Asquith and Booth (1968), Wagner (1969) and Mukerjee and Ghosh (1963). However it is well known in practice that it is much more effective to mix the dry dye powder with urea and then paste with water before dilution than to add urea to an unsatisfactory dye solution. This effect indicates some of the complexities of the role of urea as a disaggregating agent. Gilchrist (1971) has shown that in very strong urea solution anionic dyes form an adduct with urea which does not readily hydrolyse on dilution. In strong urea solutions dye-urea adducts will crystallize out extensively. Thus the behaviour of urea in disaggregating dye molecules in aqueous solution' is partly determined by the history of the solution itself. In many cases it is possible to obtain a precipitated dye-urea adduct from a 20% urea solution in water at 25°C at a general dye concentration which would give a clear solution in the absence of urea. It could be argued that under these circumstances urea is acting as an aggregating rather than a disaggregating agent. The effect has been put to useful purpose in a cold dyeing method for wool. (Lewis and Seltzer, 1968). This urea may be used to produce either the effect of aggregation or disaggregation as required.

Dimethylformamide is also capable of forming an adduct with anionic dyes but without the complications which may be caused in the case of urea. As a highly polar solvent dimethylformide is extremely useful in the purification of dyes since it may be used to free technical dye powders of electrolytes which it will not dissolve.

V   REFERENCES

Alexander, P. and Stacey, K. A. (1952) *Proc. Roy. Soc.*, **A212**, 274.
Asquith, R. S. and Booth, A. K. (1968) *J. Soc. Dyers and Colourists*, **86**, 393.
Aston, E. (1970) University of Leeds, unpublished work.
Bechthold, M. F. and Newton, R. F. (1940) *J. Amer. Chem. Soc.*, **62**, 1390.
Blears, D. J. and Danyluk, S. S. (1966) *J. Amer. Chem. Soc.*, **88**, 1084.
Burge, D. E. (1963) *J. Phys. Chem.* **67**, 2590.
Chadwick, C. S. and Neale, S. M. (1958) *J. Polymer Sci.*, **28**, 355.
Coates, E. (1969) *J. Soc. Dyers and Colourists*, **85**, 355.
Craven, B. R. and Datyner, A. (1967) *J. Soc. Dyers and Colourists*, **83**, 41.
Craven, B. R. and Datyner, A. (1964) Proc. 4th Internat. Congress on Surface Activity, Brussels.
Datyner, A. and Delaney, M. J. (1971) *J. Soc. Dyers and Colourists*, **87**, 263.
Friedrich, K. and Stammbach, K. (1964) *J. Chromatog.*, **16**, 22.
Friedrich, K. and Stammbach, K. (1968) *J. Chromatog.*, **34**, 351.
Gilchrist, A. K. (1971) Ph.D. Thesis. Bradford University.
Hague, D. N., Henshaw, J. S., John V. A., Pooley M. J. (1971) *Nature*, **229**, 190.
Hillson, P. J. and McKay, R. B. (1965) *Trans. Far. Soc.*, **61**, 374.

Holtzer, A. and Emerson, M. F. (1969) *J. Phys. Chem.*, **73**, 36.
Kissa, E. (1969) *Text Res. J.* **39**, 734.
Lamm, M. E. and Neville, D. M. (1965) *J. Phys. Chem.*, **69**, 3872.
Lenher, S. and Smith, J. E. (1935) *Ind. Eng. Chem.*, **27**, 20.
Lewis, D. M. and Seltzer, I. (1968) *J. Soc. Dyers and Colourists*, **84**, 501.
Milicevic, B. and Eigenmann, G. (1964) *Helv. Chem. Act*, **47**, 1039.
Mukerjee, P. and Ghosh, A. K. (1963) *J. Phys. Chem.*, **67**, 63.
Nemoto, Y. and Imai, T. (1959) *J. Chem. Soc. Japan* (*Ind. Chem. Soc.*), **62**, 1268.
Rattee, I. D. (1953) *J. Soc. Dyers and Colourists*, **69**, 288.
Robinson, C. and Moilliet, J. L. (1934) *Proc. Roy. Soc.*, **A143**, 630.
Robinson, C. (1935) *Proc. Roy. Soc.*, **A148**, 681.
Rosen, F. D. and Davies, L. (1953) *Rev. Sci. Instr.*, **24**, 349.
Schwenn, G. (1933) *Melliand Textilber*, **14**, 22.
Sirianni, A. F. and Puddington, I. E. (1956) *Canad. J. Chem.*, **33**, 755.
Valkó, E. I. (1937) *Kolloidchemische Grundlagen der Textilveredlung*, Springer Verlag, Berlin.
Valkó, E. I. (1945) I. G. Farben Technical Report. Captured German microfilm FD281/51, Frames 1149–1162.
Valkó, E. I. (1949) *J. Soc. Dyers and Colourists*, **65**, 217.
Valkó, E. I. (1972) Review of Progress in Colouration, **3**, 50.
Wagner, D. (1969) *Textil. Praxis*, **24**, 310.
West, N. and Pearce, S. (1965) *J. Phys. Chem.*, **69**, 1894.
Wurzschmitt, B. (1950) *Z. Anal. Chim.*, **130**, 1105.

Chapter 5

The adsorption of dyes by proteins and polyamides

I PROTEIN AND POLYAMIDE SUBSTRATES

In practical terms protein substrates take the form of wool and silk in the textile field but interactions with many other proteins are of importance in the fields of enzymology, bacteriology, biochemistry and food science etc. The related polyamide substrates are entirely man-made and fall under the general classification, nylons.

A Chemical structure

The broad chemical composition of proteins was established many years ago as being α-amino carboxylic acids in linear polymers.

$$NH_3^+\!\!-\!\!CH\!-\!\!CO\!-\!\!NH\!-\!\!CH\!-\!\!CO\!-\!\!NH\!-\!\!CH\!-\!\!COO^-$$
$$|\qquad\quad|\qquad\quad|$$
$$R\qquad\quad R^i\qquad\quad R$$
$$(\ )_n$$

α-amino acid	type (R)	structure
Glycine	Non-polar	$NH_2.CH_2.COOH$
Alanine		$CH_3.CH(NH_2)COOH$
Phenylalanine		$C_6H_5CH_2.CH(NH_2)COOH$
Valine		$(CH_3)_2CH.CH(NH_2)COOH$
Leucine		$(CH_3)_2CHCH_2CH(NH_2)COOH$
Proline		
Oxyproline	Hydroxy	
Serine		$HO.CH_2.CH(NH_2)COOH$
Threonine		$CH_3.CH(OH)CH(NH_2)COOH$
Tyrosine		
Methionine	Thio-	$CH_3.S.CH_2CH_2CH(NH_2)COOH$
Cystine		$HOOC.CH(NH_2)\text{-}CH_2\text{-}S$ $\quad\quad\quad\quad\quad\quad\quad\quad |$ $\quad\quad\quad\quad\quad S\text{-}CH_2CH(NH_2)COOH$
Arginine	Basic	$NH_2.C(:NH)NH(CH_2)_3CH(NH_2)COOH$
Lysine		$NH_2(CH_2)_4CH(NH_2)COOH$
Tryptophane		
Histidine		
Aspartic acid	Acidic	$HOOC.CH_2CH(NH_2)COOH$
Glutamic acid		$HOOC\ \ CH_2CH_2CH(NH_2)COOH$
Hydroxyglutamic acid		$HOOC.CH(OH)CH_2CH(NH_2)COOH$

The side chains R may have a wide variety of forms to include some twenty different amino acids. It is the variations on the theme of the group R and the molecular weight (chain length) which determine the physical as well as the chemical properties of the protein. The main α-amino acids involved in proteins are classificable in terms of the general properties of the side chain as shown in Table 5.1. In normal proteins many of the carboxylic side chains are present as amides.

The polyamides present (at least nominally) a much simpler picture. They do not normally possess side chains and are not cross-linked as will occur with a protein containing the amino acid cystine which can participate in two polymer chains at the same time to give a disulphide bond between them. Polyamides are of two chemical types according to how they are prepared,

acid/base condensates, e.g.

$$n \left[NH_2(CH_2)_6NH_2 + COOH(CH_2)_4COOH \right] \longrightarrow \left[NH(CH_2)_6NHCO(CH_2)_4CO \right]_{n-1}$$

This nylon is Nylon 6,6 signifying the six carbon atoms in the monomers. If adipic acid is replaced by sebacic acid, Nylon 6,10 is formed.

self condensates, e.g.

$$\cdot \quad n \left[NH_2(CH_2)_5COOH \right] \longrightarrow \left[NH_2(CH_2)_5CO \right]_{n-2} \quad \cdot$$

This nylon is Nylon 6 signifying the six carbon atoms but only a single number is needed since assymmetry is not possible.

The above is of course a highly idealized presentation of the structure of nylons. Numerous side reactions can occur during polymerization leading to the formation of pyrroles and per-acids. In addition monocarboxylic acids are normally added to control the degree of polymerization so that the chemical composition of the nylon is not simply determined by the monomer.

B   Physical structure

The physical structure of proteins is markedly affected by the arrangement of amino acids (and hence the side chains) in the polymer. In addition the acidic side chains are weak acids so that the coulombic effects within the polymer and between polymer chains are affected by the pH conditions. In the presence of water a significant effect is due to intramolecular hydrophobic interactions between non-polar residues in the polymer chain with a consequent structuring. Additionally hydrogen bonding plays an important role in this connection. It is known in some cases that adsorbed molecules will modify the physical configuration of protein molecules thus producing chemical changes, e.g. deactivation of enzymes.

In the high molecular weight insoluble proteins there is present both amorphous and crystalline character. It would not be correct to make too great a distinction between these since there is almost certainly a continuum of order in the polymer ranging from virtual randomness to virtual crystallinity. This effect as well as configurational factors determine the accessibility of components of the polymer to reagents such as dyes and also chemical modifying agents. Insoluble proteins also exhibit cellular structure deriving from their natural growth.

It is not appropriate here to consider in great detail what is known about the physical structure of proteins. This extensively documented elsewhere. The essential things to be remembered are that

(1) the structure is consequent upon the chemical composition
(2) it will be affected by reagents, water, acids and bases
(3) as far as dye adsorption is concerned the important parameter is likely to be the physical configuration of the protein at the adsorption site.

Further information regarding the general level of knowledge about protein configuration may be gleaned from the general literature.

The physical structure of polyamides is more readily defined than that of proteins since its study is largely confined to the two polyamides of major importance, i.e. Nylon 66 and Nylon 6.

II INTERACTIONS WITH WATER

These play an important part in relation to dye sorption phenomena both from the point of view of the chemical consequences of the adsorption and that of the effect of chemical modification on the adsorption. In addition the interaction of proteins and polyamides with water is of clear importance in aqueous dyebaths.

A Physical Binding

The binding of water by functional groups has been studied thoroughly by Watt and Leeder (1968) who produced adsorption isotherms of water on wool samples in which specific groups were blocked by chemical modification. The results were used to prepare differential water isotherms relating to particular groups in the wool and the following conclusions were drawn with regard to water binding capacities.

The value for protonated amino groups is generally supported by other work on small molecules as is the value for interaction with the peptide group.

Functional group	Water binding capacity (group)
ionized carboxylic acid	2 molecules
protonated amino group	3 molecules
hydroxy groups	2 molecules
peptide group (–NH–CO–)	1 molecule

(Breuer (1972a); Fraenkel and Kim (1966); Kennerley and Tyrrell (1968); Mellon *et al.* (1948).) In the latter connection Breuer (1964a) has found that on human hair phenol replaces water in a 1:1 correspondence at the peptide bond.

Watt and Leeder's data for peptide group interaction was obtained by difference. That this is probably a fortunately chosen method is indicated by the fact that water uptake is markedly affected by protein and polyamide structure. This means that simple analytically based calculation could be misleading.

Breuer has reviewed the water binding of keratin fibres (Breuer 1972b) and compared it with data from other proteins (e.g. β lactoglobulin, Zein, egg albumen etc). Water binding data of the various proteins could all be interpreted by postulating a direct water-peptide group interaction. There are wide differences in behaviour between the proteins but there is a strong indication that water uptake decreases with increasing helical content. The effect of physical structure is very clearly to be seen with polyamides from the work of Puffr and Sebenda (1967). They expressed water adsorption capacity of Nylon as a function of the degree of crystallinity of the polymer. From their data one is led to the conclusion that completely amorphous nylon would take up 1·5 mole of water per peptide bond. In both proteins and polyamides the evidence suggests that it is the water binding to peptide which is particularly affected by the physical structure of the polymer.

Hydrophobic hydration, i.e. the quantity of water immobilized by the protein or polyamide due to structuring around non-polar residues, is not likely to account for much of the water binding. It has been shown that with detergents the ionic head groups disturb the water structure at a distance of up to six carbons atoms along the non-polar chain (Clifford and Pethica, 1965); Corkill *et al.*, (1967)). This means to say that in a charged polymer regions with hydrophobic character must exist at quite a large distance from charged groups before the hydrophobic hydration can develop. Undoubtedly such regions exist but the limitation specified will considerably restrict their number.

So far the effect of the protein on the water binding has been considered. However the binding of water can bring about changes in the protein. The secondary and tertiary structure of the protein is to a degree determined by intra-molecular hydrogen bonding. Binding water molecules introduces a competitive influence so that the configuration may change. It is well established that in semi-crystalline polymers the melting point is greatly influenced by the interaction between the polymer and the solvent (Flory, 1953). The melting point of keratin has been shown by DTA (Haley and Snaith, 1967) to decrease from 220°C to 140°C with increasing water uptake. Bixon and Lifson (1966) have developed a quantitative theory which interrelates the helicity or intra-molecular bonding in a polymer with the solvent binding. This suggests a number of important conclusions.

    (a) that the helical content of a protein will depend at any temperature on the water content

    (b) that the helical content may have several values consistent with a given temperature and water content

    (c) that the hysteresis apparent in hydration/dehydration/rehydration curves on proteins reflects the existence of valid alternative states and *not* metastable states.

In addition to amorphous and α-helical states a protein may also adopt a linear β-configuration with yet another pattern of hydration behaviour.

Clearly the process of water uptake and, in aqueous dyebaths, of water saturation is very complex leading to irreversible modifications. The situation is further complicated by changes due to temperature. Polyamides may present a simpler picture than proteins in this connection but temperature dependent transitions are known. As a consequence the application of classical thermodynamics to such a system, which is never in a state of truly reversible equilibrium, is hazardous except when used in a broad sense.

B   Chemical interactions with water

As well as physical changes due to interaction with water, chemical changes also occur. These are due mainly to hydrolysis but other effects are important.

In both wool and other natural proteins and polyamides a significant proportion of acid side chains are in the amide form ($-CONH_2$). Consequently quite mild treatments lead to the generation of carboxyl groups and the liberation of ammonia into the dyebath. This is generally accepted in the case of protein studies at higher temperatures but is less generally realized in the case of polyamides. Marfell (1971) has shown that Nylon 6 yields considerable quantities of ammonia in dyeing treatments and that the quantity depends upon time, pH and the anions present. The rate determining factor would be expected to be the internal rather than the external pH and this generally appears to be true.

The imido or peptide links in proteins and polyamides are also susceptible to hydrolysis under acid and alkaline conditions in particular. The stability of the peptide link in proteins depends upon the molecular configuration and the vicinal amino acids. Long has shown significant differences in susceptibility to acid hydrolysis on the part of simple di- and tri-peptides which he ascribes to steric or probability factors. Peptide hydrolysis will be considered further in relation to adsorption interactions. In the case of polyamides the hydrolysis of imido groups is particularly associated with the phenomenon of "over-dyeing", i.e. the adsorption of dye anions by the fibre in excess of electrostatic saturation. This condition is associated with a rapid fall in internal pH and a consequently marked increase in proton catalysed hydrolysis as shown in Fig. 5.1.

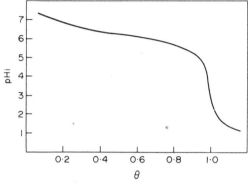

Fig. 5.1.   The relationship between pH_i and θ is a typical case.

Figure 5.1 shows the calculated internal pH/θ relationship for a typical Nylon 6 polymer in a dyebath containing pure free dye acid. In fact the internal pH would not show so marked a fall at $\theta = 1$ because the additional protons would be taken up by imido groups with subsequent hydrolysis. This kind of situation is particularly relevant when sulphonated reactive dyes are applied to nylon since each dye molecule reacting will not only bring additional negative charges, it will also "remove" a positive charge by reaction with an amino group. Thus for a monosulphonated reactive dye the above relationship would be seen at half of the values of θ shown, i.e. an internal pH approaching 1 would be seen when θ approached 0·5. The hydrolysis of Nylon 6 by free dye acids has been studied extensively by White (1968) and also Marfell (1971). They found that imide hydrolysis leading to the appearance of new amine end groups was dependent entirely upon the internal pH effect and was not affected by dye structure. Hydrolysis commenced precisely at the point where θ reached unity in all cases where the sulphonic acid groups of the dye were significantly strong to remain ionized under the low pH conditions existing at that point.

In the case of wool and other proteins containing cystine residues hydrolysis may also occur with a fission of disulphide bonds.

$$-CH_2-S-S-CH_2- \quad \xrightarrow{H_2O} \quad -C\overset{\displaystyle O}{\underset{\displaystyle H}{\diagdown}} + H_2S + HS-CH_2-$$

The rupture of the disulphide bonds particularly at high temperatures enables the wool to achieve rapidly new configurations of high stability, so called "set". Thus in this case both chemical and physical changes occur and in addition hydrogen sulphide is liberated. The reaction occurs quite slowly at pH values below 5 but in weakly alkaline conditions it proceeds quite rapidly (Alexander and Hudson, 1954).

Another aspect of the treatment of high molecular weight proteins with water is the presence in them of low molecular weight soluble fractions. Zahn has termed these "wool gelatin". They are of unknown origin but are always present. Their composition is much the same as that of the parent protein.

Thus any aqueous dyeing experiment involving wool proteins or polyamides must be expected to produce a dyebath solution containing soluble products of protein or polyamide hydrolysis, i.e. ammonia, hydrogen sulphide and low molecular weight protein fractions, as well as cause permanent damage such as setting in the case of wool. In the case of polyamides there are still further complications to be considered.

Marfell has demonstrated that Nylon 6 polymers yield in addition to ammonia on aqueous treatment, significant amounts of ε-amino per-caproic acid indicating that an oxidation mechanism is in operation. The quantity of the per-acid liberated is a function of the conditions of treatment and it clearly does not arise from a simple included impurity. Nylon 66 does not produce per-acids but shows a marked tendency to lose amino groups with the formation of pyrroles particularly in dry heating even for short times but also on heating in weakly alkaline solutions (Marek and Lerch, 1965).

The problem in studying dye sorption on proteins and polyamides due to the reactivity of the substrate and consequent irreversible changes means that absolute measurements cannot be made. In most studies problems due to changes in the polymer are compounded by the unknown composition of the dyebath. Clearly if the dyebath liquor contains soluble proteins, ammonia or dissolved fractions from a polyamide it will be comparable with the substrate in its capacity to bind dye or acids etc. This can be shown particularly clearly with experiments in which free dye acids are "equilibrated" with proteins. The free dye sulphonic acids are strong acids and freshly prepared in distilled water give solutions of the theoretical pH, i.e. the equivalent concentrations of the anions and protons are the same. If dye liquors produced using such solutions are studied it is found that at the end of a dyeing experiment a tremendous discrepancy can exist.

Taking data from Lemin and Vickerstaff (1947) provides the examples of monosulphonated dyes shown in Table 5.2.

The last column, pD, is not due to Lemin and Vickerstaff but is $-\log$ (Dye concentration) and shows the extent and variability of the discrepancy. Similar discrepancies are shown by data due to Nursten following studies of dye uptake by collagen at low temperatures. In the case of Lemin and

TABLE 5.2. Equilibrium data for the titration of wool with monobasic
acid dyes (Lemin and Vickerstaff, 1947).

Dye	pH	Dye concentration (m)	pD
Metanil Yellow YK	4·92	$5·55 \times 10^{-4}$	3·26
Coomassie Red G	5·62	$2·72 \times 10^{-4}$	3·56
Solway Blue R	3·45	$1·77 \times 10^{-4}$	3·75
Naphthalene Orange G	4·60	$1·29 \times 10^{-3}$	2·89

Vickerstaff the errors arise from the very precautions which they took to minimise damage to the wool! The low pD values signify that cations other than protons are present in the dyebath at a concentration level comparable with the protons themselves thus giving the dye an apparently higher affinity for the aqueous phase than is in fact the case. In their experimental work Lemin and Vickerstaff used an experimental technique due to Gilbert (1944). This was developed to achieve minimal wool damage in dyeing equilibrium studies. A sample of wool is dyed very quickly under conditions giving uniformity and penetration with as little damage as possible. A portion of this dyed wool is then placed in a Gilbert desorption cell shown schematically in Fig. 5.2.

The cell placed in a thermostatted bath contains water or a dye solution at some concentration such that the tendency is for dye to be desorbed from the wool. Using the intermittent pressure the liquor is circulated through the wool sample. After a short time the sample is replaced and the process repeated. After several repetitions of the sequence sufficient dye has been desorbed from the wool samples to give an "equilibrium" with the original dyed wool sample. Thus the dye liquor at this point relates to the virtually undamaged wool. However as the pH/pD relationships show, these precautions result in a build up of soluble wool extracts in the liquor giving erroneous or deceptive results.

The alternative approach is to use the same substrate sample and repeatedly change the dye liquor until no change occurs. Thus the dye liquor at "equilibrium" is a fresh solution. The substrate has been in this case subjected to prolonged treatment, but this procedure is on the whole to be preferred. The very fact that an equilibrium can be obtained suggests that the substrate has been brought to some steady condition. Certainly the amount of amonia and soluble components which can be extracted from wool and polyamides is limited. The main recommendation for the procedure is its high reproducibility which supports the views on which the method is based.

The two methods of approach yield different results and there have been very few comparisons. Two titration curves showing Nylon 6 in the presence of hydrochloric acid have been produced by Chantrey and Rattee (1969) and are shown in Fig. 5.2.

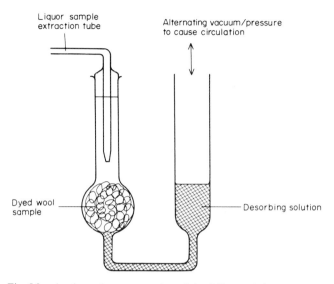

Fig. 5.2. A schematic representation of the Gilbert cell (Gilbert, 1944).

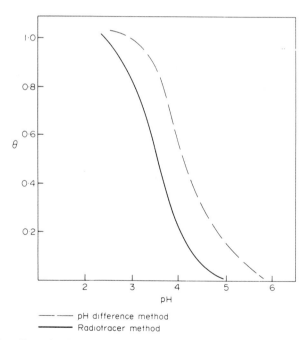

Fig. 5.3. The effect of soluble degradation products on the titration of Nylon 6 with hydrochloric acid (Chantrey and Rattee, 1969).

The full line shows the results obtained by the standard technique of immersion of the nylon sample in an acid solution and measuring the acid uptake by changes in the pH. The broken line shows those obtained when the repeated liquor change method is used and the value of θ determined by using CI^{36} radioactive tracer, extraction and counting. The discrepancy between the two reveals the same lack of correspondence between pH and pCl as in the cases with wool and collagen with pH and pD. In the case of nylon there does appear to be a point at which extracted matter ceases to appear so that an equilibrium state is attainable in at least an experimental sense.

The loss of soluble matter from wool protein is sufficiently great and, as has been shown by Marfell, varies with pH to a sufficient degree to make published data highly suspect. Lister and Peters (1954) in their examination of Naphthalene Orange G adsorption by wool estimated the content of nitrogeneous matter in the dyebath and calculated their results on the basis of its being entirely ammonia. This is quite wrong but would tend to make their results more true and indicative of dye binding by wool than other data.

III   THE BINDING OF ACIDS AND DYES BY SOLUBLE PROTEINS

It is convenient to consider soluble and insoluble proteins separately in this connection because equilibria can be established sufficiently quickly in adsorption interactions involving the former to avoid the problems arising due to degradation in many cases. In addition much more detailed knowledge is available about the protein structure in most instances. Consequently knowledge acquired in the study of soluble proteins may be applied to the study of insoluble proteins more often than the other way about.

A   Simple acid and base binding

The hydrophilic behaviour of a protein is related to its net polar character. Since the latter arises from the polar side chains the solubility properties of a protein would be expected to be pH dependent showing a minimum in circumstances where there is no net charge on the protein (the so called isoionic state). Due to the contribution of hydrophobic interactions to the properties of the protein and the contribution of water structure to the ionization of weak acids and bases, the effect of pH would be expected to be complicated by temperature effects. Thus it may be more correct to consider proteins in solution rather than "soluble proteins" in connection with adsorption phenomena.

The simplest case is that of acid/base binding where there is no anion affinity or (on the alkaline side) cation affinity other than that pertaining to the proton. This situation is reflected in the so-called titration curve of proteins obtainable with hydrochloric acid and caustic soda since neither chloride nor sodium ions appear to exhibit affinity. A theoretical titration curve can be constructed for a model protein simply from a knowledge of the molar concentrations of acidic or basic amino acid components and their respective pK values. Considering a model protein with the following composition,

Aspartic acid	40 eq mole^{-1}	pK 3·65
Glutamic acid	40 eq mole^{-1}	pK 4·25
Lysine	30 eq mole^{-1}	pK 8·90
Arginine	20 eq mole^{-1}	pK 13·20

the net charge can be calculated from

$$\text{Net charge} = \frac{H^+[\text{Lysine}]}{H^+ + K_{\text{Ly}}} + \frac{H^+[\text{Arginine}]}{H^+ + K_{\text{Ar}}}$$
$$- \frac{K_{\text{As}}[\text{Aspartic}]}{H^+ + K_{\text{As}}} - \frac{K_{\text{Gl}}[\text{Glutamic}]}{H^+ + K_{\text{Gl}}} \tag{5.1}$$

and the results are shown in Figure 5.4.

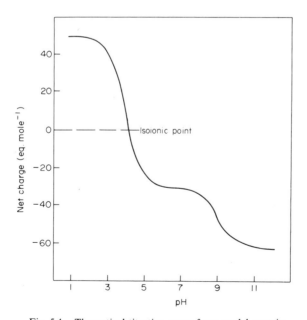

Fig. 5.4.   Theoretical titration curve for a model protein.

The isoionic point for this model protein is 4.2. In general the titration curves of proteins in solution follow the same general form. In many cases however the proteins have greatly reduced solubility as the carboxyl groups are back titrated and the charge falls, leading to precipitation. The actual determination of the isoionic point is not possible from the titration curve however since the measured parameters are the pH of the external solution and the quantity of acid bound. The isoionic point can be calculated only approximately using the pK_a and pK_b values for the constituent amino acids because these values are frequently different in the protein polymer. The isoionic point must be defined operationally as a consequence. Various definitions have been proposed but it is generally accepted that it is the pH at which no net migration of the protein is observed when the solution is subject to an electrical field (e.g. electrophoresis).

The anomalous dissociation behaviour of amino acids in proteins derives from inductive effects, i.e. the effects of condensation with other amino acids. Examples are given in Table 5.3 of substituent effects with glycine.

TABLE 5.3. Substituent effects on the pK_a and pK_b of glycines.

pK_a ($COO^- + H^+ = COOH$)		pK_b ($NH_3^+ = NH_2 + H_+$)	
Glycine	2·34	Glycine	9·60
Glycyl glycine	3·12	Hexaglycine	7·60
Aspariginyl glycine	2·90	Glycyl leucine	8·29
Glycyl glycyl glycine	3·26	Glycyl aspartic acid	8·60

An additional factor is the interaction of near neighbours. Near to an ionised acid group a high concentration of protons will exist which will have the tendency to back titrate neighbouring acid groups. Similarly there will be an increased tendency to protonate neighbouring basic groups. Thus proximity of oppositely charged groups tends to strengthen both while the opposite effect will hold when there is proximity of similar groups. The existence of this effect has been demonstrated by the work of Alfrey and Pinner (1957). Such electrostatic effects operate over significantly greater distances than do inductive effects and in a complex polymer are likely to overlap and interact. A comparison between inductive and electrostatic effects has been made by Cohn and Edsall (1943) as shown in Table 5.4.

TABLE 5.4. Inductive and electrostatic effects on pK_a values.

Acid	pK_a	Acid	pK_a
Acetic acid	4·747	Valeric acid	4·821
α-Bromoacetic acid	2·845	δ–Bromovaleric acid	4·711
α-Aminoacetic acid	2·300	δ–Aminovaleric acid	4·270

Not only is the positive group in the α-position more effective than the α-Bromo group, the effect of the latter is largely suppressed in the δ-position while that of the amino group is still considerable.

The theoretical treatment of the dissociation of polyelectrolytes is complex and has attracted considerable attention. This has been reviewed by Steinhardt and Beychok (1964). The application of the theoretical treatments to experimental titration curves requires the introduction of a number of factors not connected with acid-base equilibria such as steric effects which may isolate a potentially acidic or basic group in a hydrophobic region of the protein or cause such a group to exist in a region of abnormal dielectric constant. The situation is of such obvious complexity that it is evident why simple application of Donnan membrane theory, for example, provides only an approximate description of experimental titration curves.

The structuring effects of acid and base binding by proteins is of considerable importance. The ionization of a carboxyl group will have a significant effect both on water binding and on internal electrostatic bonding factors. In its turn structural change will effect acid or base binding. Two examples may be cited. Ferrihemoglobin shows a rapidly reversible conformational change at pH 4·3 which gives a large increase in accessible acid binding groups. Polyadenylic acid at pH > 7 is in a flexible single stranded coil configuration while at pH < 5 it is in a rigid stranded coil.

B   The binding of dyes and related substances

In the general field of proteins in solution there is a tremendous variety of different substrates for study. The selection of a substrate for examination arises very often because of the biological significance of the protein or a possible interaction rather than because any attempt has been made to systematize the available knowledge. Thus serum albumen, which is the vehicle whereby a wide variety of substances including toxins are transported in the blood stream, has been studied extensively. Interactions between dyes and dissolved proteins are generally "many site" interactions but there are several instances of highly specific interactions which lead to the formation of molecular complexes and these have been the subject of special study. Dye binding by proteins in solution is generally accompanied by an entropy gain consistent with hydrophobic binding (Klotz, 1953; Karush, 1950). Constraints on the configuration of the dye molecule in the adsorbed state have been observed, e.g. Auramine 0 is non-fluorescent in the unbound state but exhibits strong yellow-green fluorescence when bound to desoxyribose nucleic acid (Oster, 1950). It is convenient to consider the state of knowledge in terms of particular studies and to separate the general "many site" interactions from the highly specific interactions. It is then possible to arrive at overall conclusions with regard to the operating factors.

1   *Binding at multiple sites*

This occurs in two ways. There is the expected dye–site interaction and also a phenomenon known as "*stacking*" which involves dye–dye as well as dye–site interactions. The latter shows itself as a form of surface heterogeneity and it is best considered first.

Dyes have been used for the purpose of histological staining for many decades. A phenomenon whereby a single pure dye can produce two coloured stains on different parts of a section was observed by Erlich and he termed the effect *metachromacy*. While such an effect might be expected from a consideration of the contribution of dispersion forces to the binding it was noticed that the changes in colour particularly with Acridine Orange and Methylene Blue are comparable with those produced by aggregation of the dye in solution. Michaelis (1947) suggested that meta-chromatic dyes may in fact be aggregated on surfaces, i.e. be involved in dye–dye interactions. The aggregation of Acridine Orange is studied particularly easily by spectrophotometric means since in dilute (unaggregated) solution the dye has a main absorption at 492 nm. As the concentration is increased aggregation leads to a new absorption at 464 nm.

Zanker (1952) has shown by spectrophotometric methods that Acridine Orange molecules in solution have considerable mutual affinity resulting in a partial molar free energy of dimerisation of 5·7 k.cal/mole. This value is comparable with the free energy of dye binding by proteins and consequently simultaneous dye-binding and aggregation is not unexpected. The spectral shift due to aggregation is shown in the presence of proteins (Steiner and Beers, 1958). If to a dilute (monomeric) solution of Acridine Orange a small amount of dissolved protein as added, the aggregation absorption band is immediately apparent. This band is seen to develop as protein is added until a stage is reached where its development begins to diminish and finally the monomeric absorption band returns. The two bands observed in the presence of protein were called the Complex I ($\equiv$ aggregation band) and Complex II ($\equiv$ monomer band) by Steiner and Beers (Steiner and Beers, 1959; Beers *et al*, 1958).

When first observed it was suggested that the Complex I and II bands arose from different kinds of adsorption site on the protein leading to different colours. This is a complex theory and was soon abandoned in favour of the concept of aggregation on the polymer surface or *stacking*.

The driving force for stacking to occur is simply the fact that to the free energy gain from the dye–polymer interaction is added that from the dye–dye interaction. Thus the dye molecules will have more affinity for stacks than free sites. However the double interaction of an adsorbed dye molecule introduces a greater sensitivity to probability factors as was shown by Bradley and Wolfe (1959). They found that the polymer–dye ratio required before the Complex II absorption band developed varied considerably with the protein. It is possible to show that for purely random stacking, the proportional development of the Complex II band F is related to the sites to dye ratio S/D by

$$S/D = (1 - \sqrt{F})^{-1} \tag{5.2}$$

Real proteins do not give results which conform to this relationship which suggests that F will equal 0.5 when $S/D = 3.4$. In fact for DNA, S/D is 6 when $F = 0.5$ while for Heparin S/D has a value of 2170 at that point. Thus there is a probability term which distorts the situation. If this is expressed as K given by

$$K = \frac{P_1}{P_2} = \frac{\text{Probability of stacking}}{\text{Probability of monomeric adsorption}} \tag{5.3}$$

then a modified expression for S/D is obtained

$$(S/D) = (1 - \sqrt{F})^{-1} + \sqrt{F}(K-1)(1 + F - \sqrt{F})(1 - \sqrt{F})^{-1} \tag{5.4}$$

When there is random stacking $K = 1$ and equation (5.4) is the same as (5.2). Calculated values of K vary widely as can be seen.

$$\text{DNA} \qquad\qquad K = 1.96$$

$$\left.\begin{array}{l}\text{Polyphosphate} \\ \text{Heparin}\end{array}\right\} \quad K = 600$$

Some of the data of Bradley and Wolfe (1959) are shown in Fig. 5.5.

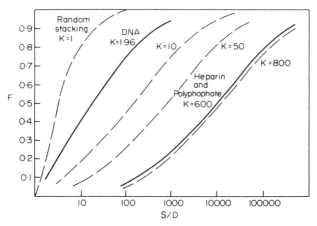

Fig. 5.5.   A comparison of observed and predicted values of F using the data of Bradley and Wolfe (1959).

There are a number of factors which would be expected to determine the stacking probability which may be itemized under two headings:

Coulombic factors: the formation of a stack involves a potential energy factor because of the charge on the dye molecules. Thus the local dielectric constant and the charge density on the fibre will be important.

Steric factors: the extent to which thermal motion of dye and substrate molecules is constrained by the dye binding will vary with circumstances and thus will affect the stacking situation. In addition the probability of forming a stack will be affected by the orientation imposed on adsorbed dye molecules by the physical configuration of the substrate. The more rigidly arranged the substrate molecule, the lower K would be expected to be.

There is considerable experimental support for the last factor. DNA which has a rigid two-stranded configuration has a low K value (1·96) while heparin and polyphosphate which exist as flexible single-stranded coils have high K values. The purine mononucleotide, polyadenylic acid, adopts two configurations according to pH. At pH values > 7 it is a single-stranded coil configuration and K is found to be high. When the pH is below 5 it adopts a rigid two-stranded helical configuration and K falls almost to the low value given by DNA.

The effects of high stacking probability have an important and interesting repercussion which has been considered by Beers and Armilei (1965). He observed that in quantitative cell staining experiments on *micrococcus lysodeikticus* and *bacillus cereus* the fraction of cells stained with acridine orange was proportional to the amount of dye bound by a given suspension of cells, i.e. the percentage saturation of the suspension was equal to the percentage of cells saturated. This unexpected phenomenon he termed "quantum binding". The situation was considered theoretically using a kinetic model involving stacks. Considering the equilibrium for monomeric adsorption,

$$AO + S_1 \overset{K_1}{=} C_1 \text{ and } [AO][S_1]/[C_1] = K_1 \tag{5.5}$$

where $[S_1]$ is the concentration of monomeric sites and $[C_1]$ is the concentration of monomerically adsorbed molecules in equilibrium with a concentration of acridine orange $[AO]$ and the equilibrium is described by a constant K_1.

The formation of stacks is described by a series of equations involving stack building by addition of an extra molecule of acridine orange. Thus

$$C_1 + AO \overset{K_2}{\rightleftharpoons} C_2 \qquad [AO][C_1]/[C_2] = K_2 \text{ for dimers} \tag{5.6}$$

$$C_2 + AO \overset{K_3}{\rightleftharpoons} C_3 \qquad [AO][C_2]/[C_3] = K_3 \text{ for trimers} \tag{5.7}$$

$$C_{n-1} + AO \overset{K_n}{\rightleftharpoons} C_n \qquad [AO][C_{n-1}]/[C_n] = K_n \tag{5.8}$$

Implicit in this treatment is the assumption that only the terminal molecules in a stack are involved in the equilibrium. This means that at any one time only a proportion of the molecules are involved in the equilibrium so that in the following example,

I II III III I IIIIII IIIIIIIII IIII I

which involves 30 molecules in stacks, only 15 are involved in the adsorption/desorption process. The situation also has an important effect on the meaning of the concentration terms, C_2, $C_3 \ldots C_n$. Since each stack has two terminal sites the "kinetic" concentration (i.e. that used in equations (5.6, 5.7 and 5.8) is double the real concentration. In computing the total concentration $\bar{C}$ from the values of C_2, $C_3 \ldots C_n$ it is necessary therefore to divide by two so that,

$$\bar{C} = C_1 + \frac{2C_2}{2} + \frac{3C_3}{2} + \frac{4C_4}{2} + \ldots \frac{nC_n}{2} \tag{5.9}$$

To proceed further it is helpful to make a simplifying assumption regarding K_2, $K_3 \ldots K_n$. Each equilibrium constant relates to the equilibrium an external concentration [AO] and the terminal position of a stack. It is reasonable to assume that this will not be affected by the size of the stack so that K_2, $K_3 \ldots K_n$ are equal and may be represented by $\bar{K}$. Thus

$$[C_1]/[C_2] = [C_2]/[C_3] = [C_3]/[C_4] = \ldots [C_{n-1}]/[C_n] = \bar{K}/[AO] \tag{5.10}$$

Using equation (5.10) $\bar{C}$ may be expressed in terms of C_1 so that writing AO/$\bar{K}$ as d;

$$\bar{C} = \frac{C_1}{2}[2 + 2d + 3d^2 + 4d^3 + \ldots nd^{n-1}] \tag{5.11}$$

Providing $r < 1$ the series converges and

$$\bar{C} = \frac{C_1}{2}\left[1 + \frac{1}{(1-d)^2}\right] \tag{5.12}$$

$$\frac{[\bar{C}]}{[C]} = \frac{1}{2}\left[1 + \frac{1}{(1-d)^2}\right] \tag{5.13}$$

The effect of this relationship can be seen by using numerical examples;

If $[\bar{C}]/[C_1] = 20$ (i.e. 95% stacking) equation 5.13 gives $d = 0.84$.

From the definition of d, $[AO] = 0.84\bar{K}$ $\tag{5.14}$

Also from (5.5), $[AO] = K_1[C_1]/[S_1]$ $\tag{5.15}$

The value of $[S_1]$ is the unoccupied site concentration, i.e. $[S] - [\bar{C}]$ where $[S]$ is the total number of sites. Applying this substitution and using the initially assumed numerical value of $[\bar{C}]/[C]$

$$[AO] = \frac{0.05K_1[\bar{C}]}{[S] - [\bar{C}]} = \frac{0.05K_1\theta}{1-\theta} \text{ where } \theta = [\bar{C}]/[S] \tag{5.16}$$

Combining (5.14) and (5.16) gives

$$\frac{K_1}{\bar{K}} = \frac{0{\cdot}84(1-\theta)}{0{\cdot}05\,\theta} \tag{5.17}$$

If a value of $\theta = 0{\cdot}01$ is set as an arbitrary value at which 95% stacking might be observed $K_1 \simeq (1{\cdot}5 \times 10^3)\bar{K}$.

Beers' studies on cell staining suggest that in his case the value of $K_1/\bar{K}$ is of the order of 10^3 so that stacked adsorption is expected to predominate. The same conclusion is reached if the situation is considered descriptively. If a stack is formed it will tend to grow in size because the proportion of molecules in it which can escape will become smaller as the number of molecules in the stack increases. Thus stacks will tend to *decrease in size* according to the value of $2/n$ where n is the number of molecules in the stack. On the other hand stacks will tend to *grow in size* at more or less the same rate irrespective of size. Consequently if there is any tendency to form a stack at all, the stacks are likely to be as large as environmental factors permit.

These conclusions are supported by Beers' results. Because of the effective molecular weight difference between cells saturated with Acridine Orange and unstained cells Beers and Armilei (1965) was able to separate the two kinds by centrifugation. To facilitate separation by enhancing molecular weight differences, Beers used polyadenylic acid as a substrate as well as two other polynucleotides. The results of density gradient ultracentrifugation are illustrated. The dye concentration was adjusted to give an overall 50% saturation of the polyadenylic acid. It can be seen that the polyadenylic acid splits into two main fractions in the presence of acridine orange approximately equal in amount. The slower moving is identical in behaviour with unstained nucleotide. The faster-moving fraction when studied at 464 nm is revealed as containing almost all of the dye. There is evidence also of precipitation of a small fraction of coloured polymer at the high density end. The colouration of about half of the nucleotide with all of the dye suggests that 50% overall saturation turns out to be 100% saturation of 50% of the nucleotide. Beers states that this was in fact the case and consistent with what could be seen microscopically with the cell staining experiments.

Alongside the development of the concept of stacking, other concepts have been developed to provide an alternative explanation of some of the experimental data. These are based on the observation of extrinsic multiple Cotton effects of Acridine Orange (and other dyes) bound to certain proteins. The Cotton effect in optical rotatory dispersion studies provides a means for the study of organic stereochemistry and conformational effects in interactions such as adsorption in certain cases (Schwarz, 1961; Hallas, 1965). Yamaoka and Resnik (1966) observed several Cotton effects in the Acridine Orange/DNA interaction. They were able to show that in this interaction at least, dye–dye interactions were not operative taking ORD curves as evidence and that the α(Complex II) bands of Beers etc. were not due to breakdown of stacks into monomeric adsorption. On another substrate (poly α L Glutamic acid) Cotton effects were observed which were not inconsistent with stacking. Current investigation into the stereochemistry of dye binding indicates that the stacking phenomenon may well be real but that some at least of the spectroscopic effects of the interaction are due to the

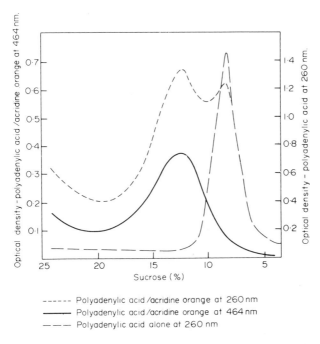

------ Polyadenylic acid/acridine orange at 260 nm
——— Polyadenylic acid/acridine orange at 464 nm
— — — Polyadenylic acid alone at 260 nm

Fig. 5.6.　Ultracentrifugation data on polyadenylic acid, Acridine Orange solutions (Beers and Armilei, 1965).

dye-protein interaction which is heterogeneous and not to dye–dye interactions alone. Beers' separation experiments and some other technical effects which will be considered later nevertheless provide strong evidence for the existence of stacking as a real phenomenon.

Other studies of multiple-site dye binding by proteins and related species in solution have tended to concentrate on the behaviour of serum albumen more than other examples of this kind of substrate. Reference has been made already to the metabolic role of this protein. It is useful to examine the investigations by Karush and Sonenburg (1949) and Eisen and Karush (1949) into the heterogeneity of the binding sites of bovine serum albumen as a typical example.

A monoazoanionic dye p-(2-hydroxy 5 methyl phenyl azo-) benzoic acid was applied at two temperatures to bovine serum albumen at pH 7·0. The results were plotted in accordance with Scatchard (1949) equations. For a heterogeneous adsorbing surface.

$$[D_f] = \sum_i K_i D_S/1 + K_i D_s (i = 1,2,\ldots S) \qquad (5.18)$$

When all the K_i's are equal

$$[D_f]/[D_s] = SK - [D_f]K \qquad (5.19)$$

which is the Scatchard equation already discussed. If the relationship for real values of $[D_f]/[D_s]$ and $[D_f]$ is not rectilinear then clearly the K_i's are not equal or there is interaction such as stacking. However the value of $[D_f]/[D_s]$ as $[D_f] \to 0$ will still be equal to $\sum_i K_i$ and the value of $[D_f]$ when $[D_f]/[D_s] \to 0$ will be equal to the site concentration S.

Karush's results are shown graphically in Fig. 5.7. $[D_f]$ signifies the concentration of bound dye expressed as moles bound dye/mole of protein.

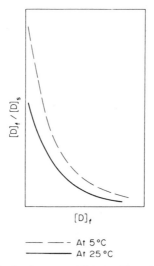

Fig. 5.7.   Dye binding by bovine serum albumen (Karush and Sonenburg, 1949).

It is instructive to examine the same results on the basis of the reciprocal plotting method which is commonly used in dye research. This is shown in Fig. 5.8.

The points are experimental values and the line is that calculated using the rectilinear regression equation (least mean squares). The correlation is excellent (0·997). The tendency for curvature can be barely detected. The data provide a good example of the insensitivity of reciprocal plotting.

Karush eliminated dye aggregation in solution as a possible cause of curvature in the Scatchard plot. Three possible factors were considered in relation to the actual dye-protein interaction.

electrostatic interaction: As dye ions build up on the surface a coulombic interaction between adsorbed molecules is to be expected. This has been considered theoretically by Scatchard (1949). The predicted effects were not

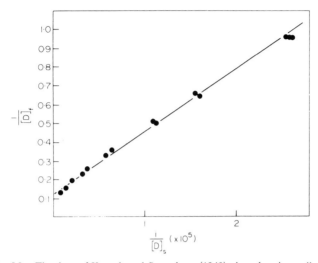

Fig. 5.8. The data of Karush and Sonenburg (1949) plotted reciprocally.

sufficiently large to account for the results and the actual effects were expected to be less than those predicted theoretically since dye anion binding would be expected to be offset by buffer anion desorption.

a gaussian distribution of sites: It was found impossible to calculate values for $[D_f]/[D_s]$ and $[D_f]$ on the basis of an assumed distribution of sites.

distribution between two kinds of site: From equation 5.18 for $i = 1$ and 2

$$\frac{[D_f]}{[D_s]} = \frac{S_1 K_1}{1 + K_1[D_s]} + \frac{S_2 K_2}{1 + K_2[D_s]} \tag{5.20}$$

Where S_1 and S_2 are the concentrations of two kinds of site and the total concentration of sites is given by

$$S = S_1 + S_2 \tag{5.21}$$

Also

$$\lim \left| \frac{[D_f]}{[D_s]} \right|_{[D_s] \to 0} = S_1 K_1 + S_2 K_2 \tag{5.22}$$

These three equations into which experimental values may be substituted provide the basis for a series of simultaneous equations which may be solved for K_1 K_2, S_1 and S_2. Karush found that excellent correlation was obtained for the assumption of two sites. The data at two temperatures showed that both kinds of site exhibited the same activation energy of adsorption but different (positive) entropies of interaction. The sites clearly differed sterically giving different water binding and hydrophobic interaction effects.

Steric factors are of considerable importance and it has been shown that whereas serum albumen is capable of binding molecules of very diverse configuration this is not true of all other proteins. Eisen and Karush (1949) describe substrates which discriminate between dye sulphonic acids and the corresponding arsonic acids for example. Some of the main features of multiple site binding of dyes by proteins have been reviewed recently by Steinhardt and Reynolds (1969).

An interesting and important effect arising from multiple site binding is the phenomenon of autolysis studied in particular by Glazer. Glazer (1969) observed that Evans Blue adsorbed onto α-chymotrypsin even at dye-:protein ratios as low as 1:1 caused rapid hydrolysis of the enzyme. In 3

(Evans Blue)

hours at pH 8·0 72% of the enzyme activity was destroyed. No other dye was found to produce as marked an effect. The isomeric Trypan Blue caused only 4% of the enzyme to be destroyed under the same conditions.

(Trypan Blue)

Similar results were obtained with the related enzyme, trypsin. No autolysis was promoted by Evans Blue with papain or Carlsberg subtilisin. Glazer concluded that the adsorption of Evans Blue on a small number of sites in α-chymotrypsin caused a stabilization of the least folded conforms of the enzyme and a consequence increase in the population of readily hy-drolysed forms. This concept is in support of the ideas of Markus (1965) who has noted the opposite effect, i.e. stabilization of proteins by bound ions. Serum albumen, for example, was stabilized by bound methyl orange ions. From his investigation Markus concluded that "the efficiency of proteolytic attack depends on the ability of the substrate protein to oscillate between a variety of conformation states". Clearly this may be increased or decreased by interactions.

Conformational changes brought about by adsorbed ions or molecules do not necessarily affect proteolysis at all. In their investigation of the effect of cresol on dye binding by α-chymotrypsin Jayaram and Rattee (1971) found a significant conformational effect with no change in protein stability.

2   Binding at single sites

This kind of dye binding by proteins is rare but when it occurs the affinity is higher than generally observed in multiple site dye–protein interactions. The complexes are 1:1 and involve the active enzyme or coenzyme sites (Brand et al., 1967) or prosthetic group binding sites (Stryer, 1965). No relationship has been found between the dye structure in these instances and that of the substrate or co-factor and it may be concluded that the binding sites offer a uniquely favourable environment for interaction with a wide variety of aromatic organic molecules. This is a conclusion with obviously far reaching biochemical or pharmacological implications.

Glazer (1969) showed that Biebrich Scarlet/α-chymotrypsin provided a typical example of single site binding. A 1:1 complex was formed with an association constant at 22°C of $1·14 \times 10^4$. The binding site was either the active site of the enzyme or such that the adsorbed dye sterically blocked the active site. A comparison of the behaviour of Biebrich Scarlet with the isomeric Crocein Scarlet suggested that the binding involved an impor-tant contribution from the naphthol component of the molecule in a weakly polar region of the protein.

Thus Biebrich Scarlet was bound more strongly than Crocein Scarlet. Hille and Koshland (1967) in a further investigation suggested that the hydroxyl group of the naphthol residue was attached to a generally hydrophobic region near to a histidine residue in the enzyme. Glazer argues that a non-polar contact is important in the dye binding because indole may be used to displace the dye. This would imply an *isosteric* competition effect. However more recent results by Jayaram and Rattee (1971) suggest that the effect of the indole may be configurational leading to an *allosteric* competition by analogy with the behaviour of cresol. The latter also breaks down the Biebrich Scarlet/α-chymotrypsin complex and it has been shown that cresol binding changes the capacity of the enzyme to bind the dye in an allosteric fashion.

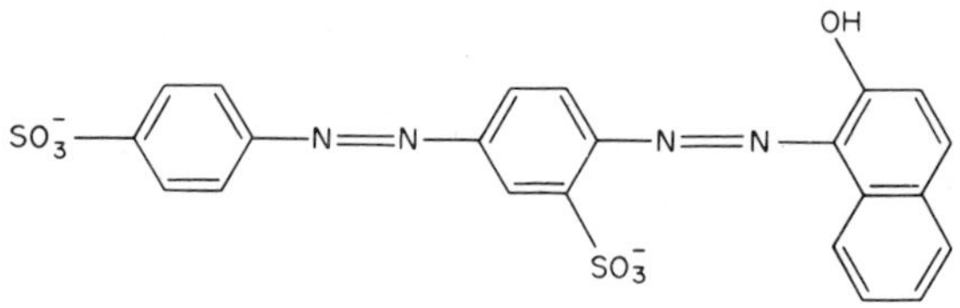

Biebrich Scarlet

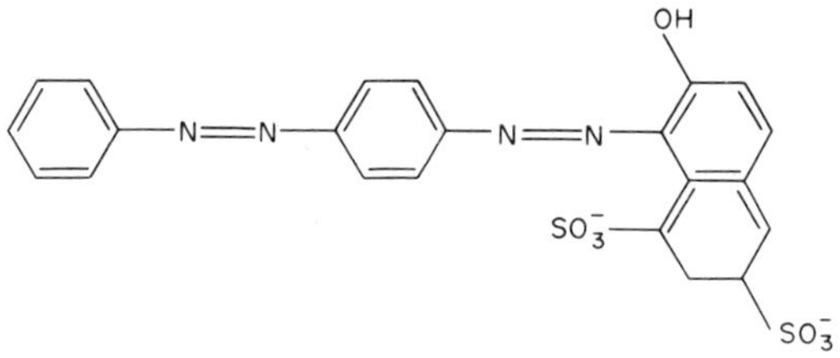

Crocein Scarlet

The very high specificity of these 1:1 interactions can be of great value in enzymological studies. Thus the spectra of proflavine/chymotrypsin and proflavine/trypsin complexes differ indicating an environmental difference. The dye discriminates between the two sites as neatly as the natural substrates and much better than many other reagents (Glazer, 1965).

IV   THE BINDING OF ACIDS AND DYES BY INSOLUBLE PROTEINS AND POLYAMIDES

The study of adsorption onto insoluble amphoteric substrates introduces the complications of a two-phase system and also longer times to achieve equilibria because of the intrusion of diffusion processes within the water swollen substrate.

In the case of adsorption from solutions onto water swollen amphoteric substrates, two equilibria rather than one have to be considered,

Dye in solution ⇌ Dye in "internal" solution ⇌ Dye adsorbed.

Experimentally studies are possible only of the equilibrium between dye in the external solution and a composite of substrate and internal solution. This has the effect of changing the nature of titration curves. In Fig. 5.9 the net charge is shown for a theoretical substrate existing as either a soluble adsorbant or an insoluble one with an internal volume of 0·067 l/kg. It can be seen that the limits of the region of zero charge are less well defined in the case of the insoluble adsorbant due to the effects of the Donnan membrane. Phenomenonlogically the isoelectric point is very difficult to determine.

Using electrophoretic methods particles or globules of the insoluble protein can be studied in media of different pH. The expected isoelectric point should be evident when no net movement of the particles or globules occurs under the potential gradient. However what is being studied is in fact not the protein as such but a composite of the protein and the water-swollen volume. If we consider for the moment a globule of a protein at its isolectric point, adsorption of a proton will create a positive charge. The second

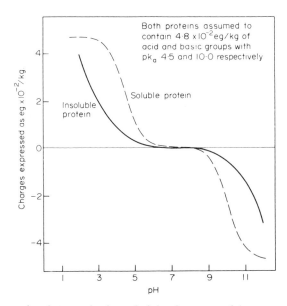

Fig. 5.9.   A comparison between the theoretical titration curves of the same protein in soluble and insoluble forms.

proton will consequently face a potential repulsion which may be strong enough to prevent adsorption. However the positive charge on the globule will attract into the internal volume negative ions so that what will occur will be the simultaneous uptake of protons and counter ions. Since their charges will neutralize one another the net charge on the water-swollen globule will not change and the system will be electrically buffered. The effect is very marked and leads to the spreading of the isoelectric point into an isoelectric region which in the case of wool, for example, covers nearly 4 pH units.

A   The theoretical treatment of the titration of proteins and polyamides

Numerous theoretical treatments have been proposed to account for titration behaviour by proteins and polyamides. Despite this there are only two approaches, that based on a specific site adsorption model put forward by Gilbert and Rideal (1944) and various modifications of Donnan membrane theory. None of the theories gives an accurate account of experimental data but it is fairly clear that none have really been put to a true test as a consequence of substrate degradation during the adsorption process. Since proper data are becoming available particularly on polyamide substrates it is of value to consider the theories in detail.

1   The Gilbert-Rideal theory

This theory assumes that there is an equal number of positive and negative charges on the fibre and that these constitute adsorption sites in a Langmuirean sense. The water-swollen fibre is taken to constitute an equipotential volume. Taking the two equilibria in turn, for all species

$$\mu_i = \mu_i^0 + RT \ln a_i \text{ where } i \text{ signifies the internal solution} \tag{5.23}$$

$$\mu_f = \mu_f^0 + RT \ln a_f \text{ where } f \text{ signifies the adsorbing surface} \tag{5.24}$$

$$\mu_s = \mu_s^0 + RT \ln a_s \text{ where } s \text{ signifies the external solution} \tag{5.25}$$

Within the equipotential volume, at equilibrium

$$\mu_i = \mu_f \tag{5.26}$$

so that

$$-\Delta\mu_{i,f}^0 = -(\mu_f^0 - \mu_i^0) = RT \ln a_f/a_i \tag{5.27}$$

There is a potential difference between the internal and external solutions so that $\mu_i \neq \mu_s$ and an allowance for the electrical potential must be made. This means that for protons

$$-\Delta\mu_{s,f}^0 = RT \ln a_f^+/a_s^+ + e\psi R/k \tag{5.28}$$

and for anions

$$-\Delta\mu_{s,f}^0 = RT \ln a_f^-/a_s^- - e\psi R/k \tag{5.29}$$

By adding equations (5.28) and (5.29) the embarrassing potential term may be eliminated to give

$$-\Delta\mu_{\pm}^0 = RT \ln a_f^+ . a_f^- - RT \ln a_s^+ . a_s \tag{5.30}$$

Gilbert and Rideal equated the activity of the adsorbed ions with the $\theta/1-\theta$ term discussed previously and the activity of the ions in solution with their concentration to give for a monobasic acid,

$$-\Delta\mu_{\pm}^0 = RT \ln \frac{\theta_+}{1-\theta_+} \cdot \frac{\theta_-}{1-\theta_-} - RT \ln [H^+]_s [X^-]_s \qquad (5.31)$$

For a simple acid solution, $[H^+]=[X^-]$ Gilbert and Rideal assumed that $\theta_+ = \theta_-$ so that

$$\frac{-\Delta\mu_{\pm}^0}{4\cdot606\,RT} = \frac{\log\theta}{1-\theta} + pHs \qquad (5.32)$$

It is useful to examine Gilbert and Rideal's argument for implicit as well as the explicit assumptions.

Firstly it is assumed that the substrate is an ideal stable substance so that only the protons and anions due to the acid are present in solution. This is now known to be quite false and since the error can be in terms of either or both ions and of an order of magnitude comparable with the concentrations of either, this is clearly a difficult matter when attempts are made to correlate theory and practice. Secondly there is no way of experimentally distinguishing between internally adsorbed and adsorbed ions. Thus the experimental θ is not the θ of equation (5.31). The θ used in equation (5.31) is $[H_f^+]/[S_f]$ for example, where S_f is the site concentration. The experimental θ is however $([H_f^+]+[H_i^+])/[S_f]$ which leads to significant differences in final values. To treat the experimental θ as being equivalent to the true θ, is to assume implicitly that ψ is zero.

Consequently it must be accepted that because of these two factors the Gilbert–Rideal theory has never been properly evaluated experimentally. The prediction of equation (5.32) is that a rectilinear relationship of unit slope should exist between pHs and $\log\theta/1-\theta$. When this prediction was tested by Vickerstaff (1954) using experimental data of Steinhardt and Harris (1940) it was found that the slope was 0·87. Peters and Speakman (1949) also pointed out that the relationship was slightly sigmoidal and not rectilinear anyway. The test relationship is a log/log form and consequently insensitive to curvature so Peters' point is significant. The error in the slope is equally so. However the source of the discrepancy has not been determined under conditions yielding no dissolved ions in the dye bath. Recent data produced by Chantrey (1971) do enable a valid test to be made. Chantrey studied the uptake of HCl[36] by Nylon 6 using the repeated liquor changing technique so that the equilibrium liquor contained only acid ions. In addition he carried out experiments using the conventional procedure. A comparison of the results obtained using the two methods is shown in Fig. 5.10.

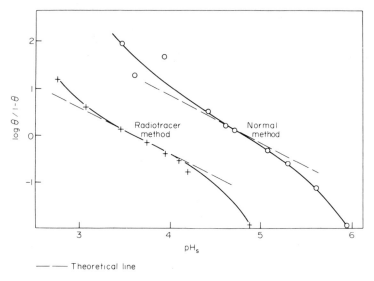

— — — Theoretical line

Fig. 5.10.   The effect of experimental method on the test of validity of the Gilbert–Rideal equation.

Clearly the two methods give very different numerical values as might be expected. In both cases the sigmoidal nature of the relationship is very clear. Thus the effect is confirmed and shown to be characteristic of the real relationship and not to be due to experimental errors. The Gilbert–Rideal theory is incorrect as a predictive model except in a very approximate sense. Vickerstaff (1954) proposed an alternative approach in which the acid anion was assumed to be taken up in accordance with a Donnan distribution. This gives the relationship

$$\frac{-\Delta\mu^0}{RT} = \ln\frac{\theta[X^-]}{1-\theta V} - \ln[H^+]_s[X^-]_s \qquad (5.33)$$

Again equating $[X^-]$ with $[H_i^+] + [H_f^+]$, Vickerstaff obtains

$$\frac{-\Delta\mu^0}{2\cdot303\,RT} - \log\frac{S_f}{V} = \log\frac{\theta^2}{1-\theta} + 2\mathrm{pH} \qquad (5.34)$$

Plotting the results of Steinhardt and Harris according to equation (5.34) gives a line with a slope of 1·4 instead of 2. Chantrey's data (1971) gives results which are no better in supporting the theory.

The value of the affinity $(-\Delta\mu^0)$ calculated using the Gilbert–Rideal equations is directly related to the value of $[H_s^+][x_s^-]$ at $\theta = 0.5$ since from equation (5.32)

$$-\Delta\mu^0 = 4\cdot606\,RT\,\mathrm{pHs} \qquad (5.35)$$

when $\theta = 0.5$ since $\theta/1-\theta = 1$

The value of the affinity reflects the experimental data at that point and at values of θ not far different from 0·5 when accurate data are used. It bears little relation to experimental results at values of $\theta > 0.75$ and < 0.25. Using uncorrected data, i.e. results obtained by the normal methods, the affinity value may express little more than a single point on the graph. It is possible that some modification of the Gilbert–Rideal model introducing special corrections for high and low θ values would prove satisfactory in a predictive sense but if that is all that is needed simplier ways of summarizing data can be found.

2 Theories based on ion exchange models

The assumptions in this approach are that the water-swollen substrate constitutes an equipotential volume and that distribution of ions is in accordance with a Donnan distribution. Using the Donnan–Guggenheim equations for equilibria at semi-permeable membranes Peters and Speakman (1949) produced the expression.

$$\text{pH}_i = \frac{(z+1)}{z}\text{pHs} + \frac{1}{z}\log\left(\frac{a}{v} + [\text{H}_s^+]\right)f_z \tag{5.36}$$

a is the anion adsorption in eq/kg, v is the internal volume in 1/kg, z is the anionic charge, f_z is the term incorporating all activity coefficients, i.e.

$$f_z = \frac{f_s}{f_i}\left|\frac{W_s}{W_i}\right|^r \tag{5.37}$$

where f_s, f_i are the activity coefficients of anion in the external and internal phases; W_s and W_i are the activity coefficients of the water in the external and internal phases; r is the ratio of partial molar volumes of the anion and water.

In fact there are no experimental data to provide a value for f_z and the apparent precision of the theory exists only at the theoretical level. In practice f_z must be treated as unity and there is no reason to suppose that this is very far out. Its contribution to the final value is in any event small.

Equation 5.36 can be derived in another simpler way. For an acid $\text{H}_z X$ the Donnan distribution equations give

$$[\text{H}_s^+]^z[\text{X}_s^-] = [\text{H}_i^+]^z[\text{X}_i^-] \tag{5.38a}$$

$$V[\text{H}_i^+] + [\text{H}_f^+] = Vz[\text{X}_i^-] \tag{5.38b}$$

The term V (the volume of the internal phase) is introduced in (5.38b) to allow $[\text{H}_f^+]$ to be expressed in g.ions/kg. If the value of $Vz[\text{X}_i^-]$ is c then

$$[\text{X}_i^-] = \frac{c}{Vz} \tag{5.39}$$

Substituting (5.39) in (5.38a) and taking logarithms,

$$z \log[H_s^+] + \log[X_s^-] = \log\frac{c}{Vz} + z \log[H_i^+] \qquad (5.40)$$

Assuming only two ions are present in solution, $[H_s^+]/z = [X_s^-]$ so

$$zH \log[H_s^+] = \log c/V + z \log[H_i^+] \qquad (5.41)$$

and

$$pH_i = \frac{zH}{z} pH_s + \frac{1}{z} \log c/V \qquad (5.42)$$

which is very similar to equation (5.36) and because $c/V \gg H_s^+$ under all conditions the values for pH_i given by the two equations are the same.

If the acid used is hydrochloric acid, wool is found to be half titrated at $pH_s = 2 \cdot 1$. Using value of V obtained from swelling data (0·3 l/kg), c is 0·4 g.eq/kg so that from equation (5.42)

$$pH_i = 4 \cdot 2 + 0 \cdot 1 = 4 \cdot 3 \qquad (5.43)$$

which is a value to be expected for carboxylic acid groups. This does not constitute any sort of proof of the validity of the Peters–Speakman theory. Firstly it is now known that the assumption that $[H_s^+] = z[X_s^-]$ is not valid in practice and secondly the Gilbert–Rideal equation, which at the mid point may be written

$$-RT \ln K = -\Delta\mu^0 = 4 \cdot 606 \, RT \, pH_s \qquad (5.44)$$

$$\text{or } pK = 2 \, pH_s$$

where K is the dissociation constant of the carboxylic acid groups, gives a pK value of 4·2 which is only marginally different.

Peters and Speakman also showed that the very different titration curves obtained using hydrochloric acid in the presence and absence of sodium chloride resolved to a single curve when the internal rather than external pH was used. This is shown in Fig. 5.11. This is what broadly speaking is predicted by the Donnan membrane theory, i.e. for a given measure of acid binding a fixed electrical potential and hence $[H_i^+]$ is involved. However it must be remembered that again uncorrected data were used in checking Peters and Speakman's theory. Significant amounts of ions would be present in solution which were neglected in calculating the internal proton concentration and the values of $[H_s^+]$ would not be truly representative of cation concentrations. When corrected data are used as in the case of Chantrey (1971) the correspondence of the internal pH plot is rather less satisfactory as can be seen in Fig. 5.12.

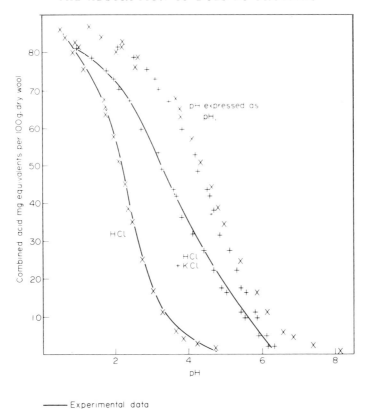

— Experimental data

Fig. 5.11. Titration data on wool expressed as a function of pH_i (Peters and Speakman, 1949).

The Donnan membrane equations give a reasonably satisfactory correspondence with experimental data providing it is corrected against decomposition products being present. Taking Chantrey's data the Donnan equations give a good correspondence with experiment with θ values <0.9 and >0.1 when hydrochloric acid alone was used, but the correspondence was observed over a much smaller range of θ values when the ionic strength was kept constant. In fact it might have been expected that things would be found the other way about and it is clear that the Donnan equations provide only a first order approximation of the real situation. The main source of error is likely to be the assumption that the chloride ion has no affinity. It probably has a very small affinity and this will make a significant difference.

Peters (1960) further developed his theories to produce a general theory which enabled affinities and titration curves to be calculated. This in effect uses the same equations as have been already discussed but the anion binding is assumed to be equal to or greater than that involved in a simple

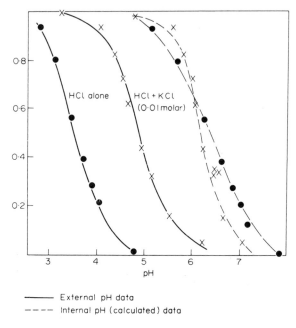

External pH data
- - - - Internal pH (calculated) data

Fig. 5.12.   Titration data on Nylon 6 (Chantrey, 1971).

Donnan distribution so that an affinity can be calculated based on the excess taken up. Considering an amphoteric substrate, e.g. wool, in equilibrium with an aqueous solution of a mono basic acid and its potassium salt (the metal anion is not important) then the ionic balance equations are

$$B^+/V + [H_i^+] + [K_i^+] = [X_i^-] + K_W[H_i^+] + A^-/V \qquad (5.45)$$

$$[H_s^+] + [K_s^+] = [X_s^-] + K_W[H_s^+] \qquad (5.46)$$

in which B and A are the concentration of the dissociated basic and acidic groups in the substrate, V is the water swollen volume of the fibre, K_W is the equilibrium constant for water inside and outside the fibre. If K_a and K_b are acidic and basic dissociation constants of the changed centres in the substrate and K_x is the binding constant of the anion then at the isoionic point

$$[H_s^+] = \frac{A}{2B}\left[1 + \frac{[X_s^-]}{K_x}\right] - \frac{1}{2} \pm \frac{1}{2}\left[\left(\frac{1-A}{B}\left[1 + \frac{[X_s^-]}{K_x}\right]\right)^2 + \frac{4K_aK_bA}{B}\right]^{1/2} \qquad (5.47)$$

where A and B are the total concentrations of acidic and basic groups. When the anion has no affinity and consequently $K_x \to \infty$, equation (5.47) becomes

$$[H_s^+] = K_a\frac{(B-A)}{2B} \pm \left[K_a^2(B-A)^2 + \frac{K_aK_bA}{B}\right]^{1/2} \qquad (5.48)$$

In the special case where $A = B$

$$H_s^+ = (K_a . K_b)/2 \qquad (5.49)$$

$$pH_s = 1/2 \, (pK_a + pK_b)$$

From equation 5.47 a value of K_x can be calculated from H_s^+ at the isoionic point and the thermodynamic affinity from the standard relation $-\Delta\mu^0 = RT \ln K_x$.

Another approach based on the ion exchange model is that of Myagkov and Pakshver (1956) which treats the system in accordance with regular ion-exchange theory. Again it is based on the calculation of appropriate dissociation or binding constants. In the original work no allowance was made for degradation of the polymer and in any event the ability of the theory to describe experimental data lies in the region of half saturation. It is not surprising that the approach is no better than that of Peters since it is based on the same assumptions.

Further developments, also based on the Donnan membrane model, are the treatments of Breuer (1964b) and of Mathieson and Whewell (1964) which are based on the general lines of the theory of polyelectrolytes. This is concerned with the regular titration of charged groups in the fibre and the energy needed to remove protons from the fibre into the solution phase against the electrostatic, osmotic and affinity forces. The equation for the mid point of the titration curve is

$$RT\,[H_s^+] = RT\,K_0(1-\alpha) + \exp\left(-\psi F - \Pi \bar{V}_H - \Delta\mu^0\right) \qquad (5.50)$$

or $\quad (pH_s)_{\mathrm{mid}} = pK_0 - \log \dfrac{(1-\alpha)}{\alpha} - \dfrac{0\cdot4343}{RT}\left(\psi F + \Pi \bar{V}_H + \Delta\mu^0\right) \qquad (5.51)$

in which pK_0 is the intrinsic pK of the groups being titrated, e.g. COO^- groups, α is the proportion dissociated at the particular pH, ψ is the surface potential, F is the Faraday, $\bar{V}_H$ partial molar volume of the hydrogen ion, Π is the osmotic pressure and $\Delta\mu^0$ is the standard partial molar free energy of adsorption. Mathieson and Whewell state that on wool the osmotic term could be ignored and that the titration curve of wool could be explained on the basis of two kinds of acid group being present (pK_1 3·85 and pK_2 4·85). No allowance was made for degradation of the wool. Barvé (1965) found that the theory did not describe the titration of nylon but he also failed to allow for fibre degradation. In any case many of his data are suspect. The approach of the Mathieson and Whewell theory is interesting but requires experimental support using corrected data. In view of its close similarity to the Donnan membrane theory and the application of that theory by Peters, it seems improbable that it will provide a more reliable description of experimental data.

The discussion of the ion exchange aspects of acid and dye-acid binding by proteins and polyamides, whether based on some specific site model of the

Gilbert–Rideal variety or one of the models based on the Donnan membrane theory evades the consideration of the different affinities of different anions. Both approaches make an implicit assumption that each ion is taken up independently. However there are good reasons to believe that this is not so.

Due to its relatively limited number of available charge centres it is relatively simple to study the dyeing of nylon in the region of its electrostatic saturation. This was first done thoroughly by R. H. Peters (1945) who showed that using dyes the titration curve of Nylon 66 could be divided into three broadly characteristic regions as shown in Fig. 5.13.

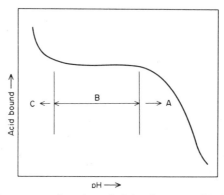

Fig. 5.13.   A schematic representation of a typical titration curve of Nylon 66 with a dye acid.

Region A represents a regular titration region with an anion of high affinity. Region B is a region of electrostatic saturation where there is broad correspondence between the acid bound and the positive charge centres in the fibre. Region C is a region in which electrostatic saturation is exceeded markedly and adsorption increases with no apparent maximum. Regions A and C are always to be seen whether an acid with high or low affinity is used. Region B is not always well defined. Dyes of high affinity and of low affinity both tend to show a weakly developed region B. A consideration of the dyeing situation giving rise to these effects is useful in the development of dyeing theory.

Considering first of all the region C, all the evidence points to this being characteristic of a situation in which the internal pH of the polymer has become sufficiently low for protanation of imido groups to occur

$$H^+ + -CONH- \rightleftharpoons -CON^+H_2-$$

This is a precondition for the onset of proton catalysed hydrolysis of the polymer as has been discussed already. R. H. Peters and O'Brian (1953) showed that in the C region in an infinite dyebath, dye was steadily taken up at the same rate as amine end groups were formed by hydrolysis but that this occurred as a consequence of the amount of dye adsorbed exceeding the electrostatic saturation level and not the other way round as suggested by Remington and Gladding (1950). Thus it is necessary to explain the driving force leading to the adsorption of the dye in excess of the electrostatic saturation level. This is clearly a non-coulombic affinity force but there are a number of other important factors also to explain.

In Fig. 5.14 are shown titration curves of Nylon 6 film produced by White (1968) using a homologous series of naphthalene azo naphthol Sulphonic acids.

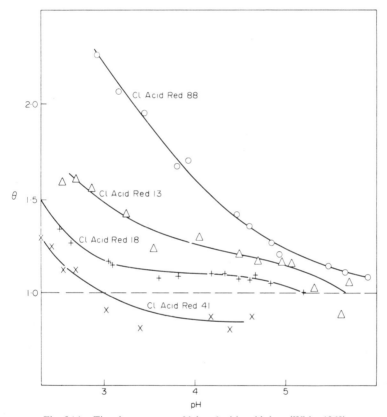

Fig. 5.14. Titration curves on Nylon 6 with acid dyes (White 1968).

The results were obtained using the method of continual dyebath replacement to avoid dissolved impurities and are, therefore, corrected data.

It can be seen that there are a number of effects associated with the varying sulphonation. Firstly the B region becomes better defined as the sulphonation increases and also the level of adsorption in that region corresponds more and more closely with the theoretical electrostatic saturation. With the mono, di and tri sulphonates it is difficult to correlate the amino

C.I. Acid red 88	4 — sulphonate
C.I. Acid red 13	4:6' — disulphonate
C.I. Acid red 18	4:6':8 — trisulphonate
C.I. Acid red 41	4:3':6':8' — tetrasulphonate

acid content of the polymer with the dyeing behaviour in any way. Nevertheless electrostatic saturation must be reached and any further dye adsorbed must lead to the development of a negative surface potential opposing further adsorption. This effect is only sufficiently marked to give a correspondence between charge factors and maximum dye uptake with the tetrasulphonate in this series. There is an effect with a lesser degree of correspondence with the trisulphonate but very little with the other two. This result suggests that there are two opposing factors in the region B. Firstly there is the free energy gain which promotes adsorption and secondly there is the potential energy gain in the surface resisting adsorption due to charge repulsion effects between anions.

The driving force for adsorption may be represented as $-\Delta\bar{\mu}_0$ given by

$$-\Delta\bar{\mu}_0 = -\Delta\mu_0 + \Pi \tag{5.52}$$

in which $-\Delta\mu^0$ is the free energy component and Π is the potential energy component. When $\Pi = \Delta\mu^0$ no adsorption takes place. Since Π will be a function of Z^2 where Z is the charge on the anion then its effect would be expected to increase as Z increases through $1 \rightarrow 2 \rightarrow 3 \rightarrow 4$. With a monosulphonate of high affinity electrostatic saturation would be expected to be exceeded very easily and this is seen with C.I. Acid Red 88.

Thus the saturation value is seen to be a consequence of opposing forces normally corresponding with the positive charge density on the polymer when dye anions are used which have moderate affinity and/or a high charge. Consistent "overdyeing" is to be expected with dyes of high affinity or low charge. This situation is a special case of dye–dye interaction on an adsorbing surface. Since the surface potential of the nylon is determined by anion binding when the electrostatic saturation level is exceeded, overdyeing is accompanied by a lowering of the internal pH leading to a protonation of imido groups at a particular level. When this occurs further dye molecules can be taken up without increasing Π and so region C of the titration curve commences.

White's results show that the coulombic interaction effect is a restraint rather than an aid to adsorption. In the region below electrostatic saturation adsorption proceeds by virtue of other binding forces but at a particular level of adsorption these are not sufficiently powerful to offset electrical potential effects.

Recently McGregor and Harris (1970) have developed a mathematical approach which makes no pretence to being a "theory" but which describes dyeing behaviour reasonably accurately. The benefit of having a mathematic analysis rather than a theory is that empirical corrections may be incorporated as necessary. Nevertheless the approach has a theoretical basis. McGregor and Harris' equation is a modification of Guggenheim's expression for ionic distribution between two phases f and s, to give

$$\frac{C_i^f}{C_i^s} = \exp\left[\frac{-\Delta\mu_i^0 + \bar{v}_i\Delta P + Z_i F\Delta\psi + RT\ln\left(f_i^f/f_i^s\right) + RT\ln\chi_i}{RT}\right] \quad (5.53)$$

This expression incorporates chemical $(\Delta\mu_i^0)$, mechanical $(\bar{v}_i\Delta P)$ and electrical $(Z_i F\Delta\psi)$ work terms, a function of activity coefficients (f_i^f/f_i^s) and a term allowing for structural effects χ_i. For simplicity an *apparent standard affinity*, $-\bar{\Delta}\mu_i^0$ is defined

$$-\bar{\Delta}\mu_i^0 = -\left[\Delta\mu_i^0 + \bar{v}_i\Delta P + RT\ln\left(f_i^f/f_i^s\right) + RT\ln\chi_i\right] \quad (5.54)$$

giving a partition coefficient K_i,

$$K_i = \exp(-\bar{\Delta}\mu_i^0/RT) \quad (5.55)$$

Also a *Donnan coefficient* λ is defined

$$\lambda = \exp\left(-F\Delta\psi/RT\right) \quad (5.56)$$

so that (5.53) may be written as

$$C_i^f = (\lambda_i)^Z K_i C_i^s \quad (5.57)$$

Electrical neutrality must exist in the solution and polymer phases so that

$$\sum_i Z_i C_i^s = 0 \quad (5.58)$$

and

$$C^+ + \sum_i Z_i C_i^f = 0 \quad (5.59)$$

where C^+ is the net positive charge on the polymer. This is calculated from the internal pH in the usual way using the dissociation constants of the acid and basic groups.

A value of λ must be found to satisfy equation (5.57) and this is found by iterative procedures on a computer. The value of C_i^f can then be calculated for any given values of K_i and C_i^s. Using this method McGregor and Harris have been able to study the expected effects of polymer composition on dyeing behaviour with many technically useful results.

V   BINDING MECHANISMS

The approach of McGregor and Harris can in fact be used for any fibre where affinity and electrostatic factors operate and being a purely empirical method it can be made to fit real data. This is the aim of quantitative theories but it does not provide any clarification of the molecular mechanism, for this other methods have to be found. For reasons which should be clear, it is not possible to produce thermodynamic data which cast a great deal of light on possible mechanisms of adsorption. The most useful alternative approach is the study of carefully designed series of compounds. A particularly useful study has been carried out by Meybeck and Galafassi (1970). A series of dyes was prepared, and their dyeing kinetics and adsorption behaviour was studied.

The group R describes a wide variety of linear and cyclic saturated and unsaturated substituents, and also a number of hydroxylated residues. Interesting differences between hydrophobic residues were observed. Hydroxyl groups considerably lessoned the substantivity of the dyes. Meybeck and Galafassi conclude that hydrophobic interactions were of great importance in the adsorption while hydrogen bonding or coulombic interactions played a secondary role.

The same conclusion has been drawn by Cockett *et al.* (1970) who examined the uptake of a series of phenols. The presence of the hydroxyl group in sterically accessible configuration was found to be of great importance but the presence of hydrophobic residues dominated the situation. Thus while sterically hindered compounds such as 2:6 xylenol showed considerably reduced affinity, the adsorption behaviour of alyklphenols was endothermic as expected for hydrophobic interactions. All of the phenolic compounds examined were able to promote swelling of the fibre and this cannot, of course, be neglected.

The degree of fibre swelling was a function of the molar adsorption of the phenol irrespective of its constitution, which suggests that the forces leading to adsorption and to swelling may be relatively unconnected. One postulate which is consistent with this view and the experimental data is that adsorption of these compounds involves a co-operative mechanism between hydrophobic interactions and hydrogen bonding, the former providing the driving force initiating the adsorption and the latter providing a secondary binding force for molecules already adsorbed and also being associated with the swelling process.

A close examination of the results of Bird and Firth (1960) tends to support the conclusion that the actual affinity forces are endothermic and arise from hydrophobic interactions. Bird and Firth examined the adsorption of a number of disperse dyes onto wool finding linear isotherms. When allowance is made for the effect of temperature on solubility of the dyes used, it is clear that partition is increasingly in favour of the fibre as the temperature rises. There has been no attempt to follow up Bird's work in this field and it does appear that it would be most interesting and useful to do so in the light of currently held ideas.

VI REFERENCES

Alexander, P. and Hudson, R. F. (1954) *Wool: its Chemistry and Physics*, Chapman and Hall, London.
Alfrey, T. and Pinner, S. H. (1957) *J. Polymer Sci.*, **23**, 533.
Barvé, S. P. (1965) Ph.D. Thesis, University of Leeds.
Beers, R. F. Hendley, D. D. and Steiner, R. F. (1958) *Nature*, **182**, 242.
Beers, R. F. and Armilei, G. (1965) *Nature*, **208**, 466.
Bird, C. L. and Firth J. (1960) *J. Text. Inst.*, **51**, T1342.
Bixon, M. and Lifson, S. (1966) *Biopolymers*, **4**, 815.
Bradley, D. F. and Wolfe, M. K. (1959) *Proc. Nat. Acad. Sci.*, (USA) 45ii, 944.
Brand, L. Gohlke, J. R. and Rao, D. S. (1967) *Biochemistry*, **6**, 3510. 68, 97, 72.
Breuer, M. M. (1964a) *J. Phys. Chem.*, **68**, 2067.
Breuer, M. M. (1964b) *Trans. Far. Soc.*, **60**, 1006.
Breuer, M. M. (1972a) *J. Coll. Interfal. Sci.*, **40**, 429.
Breuer, M. M. (1972b) *J. Cosmetic Chem.*, **23**, 447.
Chantrey, G. (1971) Ph.D. Thesis Leeds University.
Chantrey, G. and Rattee, I. D. (1969) *J. Soc. Dyers and Colourists*, **85**, 618.
Clifford, J. and Pethica, B. A. (1965) *Trans. Far. Soc.*, **60**, 1483.
Clifford, J. and Pethica, B. A. (1965) *Trans. Far. Soc.*, **61**, 182.
Cohn, E. J. and Edsall, J. T. (1943) *Proteins, Amino Acids and Peptides as Ions and Dipolar Ions*, Reinhold, New York.
Cockett, K. R. F., Kilpatrick D., Rattee I. D. and Stevens C. B. (1970) Proceedings of the 4th International Wool Textile Research Conference, (Applied Polymer Symposia, No. 18).
Corkill, J. M., Goodman, J. F. and Tate, J. R. (1967) *Trans. Far. Soc.*, **63**, 773.
Eisen, H. N. and Karush, F. (1949) *J. Amer. Chem. Soc.*, **71**, 363.
Fraenkel, G. and Kim, J. P. (1966) *J. Amer. Chem. Soc.* **88**, 4203.
Flory, P. J. (1953) *Principles of Polymer Chemistry*, Cornell U. P., New York.
Gilbert, G. A. (1944) *Proc. Roy. Soc.*, **A183**, 167.
Gilbert, G. A. and Rideal, E. K. (1944) *Proc. Roy. Soc.*, **A182**, 335.
Glazer, A. N. (1965) *Proc. Nat. Acad. Sci.*, (USA) **54**, 171.
Glazer, A. N. (1969) *Proc. Nat. Acad. Sci.*, (USA) **64**, 235.
Hallas, G. (1965) *Organic Stereochemistry*, McGraw Hill, London.
Haley, A. R. and Snaith, J. W. (1967) *Text. Res. J.*, **37**, 838.
Hille, M. B. and Koshland, D. E. (1967) *J. Amer. Chem. Soc.*, **89**, 5945.
Jayaram, R. and Rattee, I. D. (1971) *Trans. Far. Soc.*, **67**, 884.

Karush, F. (1950) *J. Amer. Chem. Soc.*, **72**, 2705.
Karush, F. and Sonenburg, M. (1949) *J. Amer. Chem. Soc.*, **71**, 1369.
Kennerley, M. and Tyrrell, H. J. V. (1968) *J. Chem. Soc.*, A, 607.
Klotz, I. M. (1953) in *The Proteins* 1st edition, ed. H. Neurath, Academic Press, London.
Lemin, D. and Vickerstaff, T. (1947) *J. Soc. Dyers and Colourists*, **63**, 405.
Lister, G. H. and Peters, L. (1954) *Discussions of the Faraday Society*, No 14.
Marek, B. and Lerch, E. (1965) *J. Soc. Dyers & Colourists*, **81**, 481.
Marfell, D. (1971) Ph.D. Thesis. Leeds University.
Markus, G. (1965) *Proc. Nat. Acad. Sci.* (USA) **54**, 253.
Mathieson, A. R. and Whewell, C. S. (1964) *J. App. Polymer Sci.*, **8**, 2029.
Mellon, E. F., Korn, A. H. and Hoover, S. R. (1948) *J. Amer. Chem. Soc.* **70**, 3040.
Meybeck, J. and Galafassi, P. (1970) Proceedings of the 4th International Wool Textile Research Conference, (Applied Polymer Symposia, No. 18).
McGregor, R. and Harris, P. W. *J. App. Polymer Sci.*, **14**, 513.
Michaelis, L. (1947) *Cold Spring Harbor Symposium on Quantitative Biology*, **12**, 131.
Myagkov, V. A. and Pakshver, A. B. (1956) *J. App. Chem.*, (USSR) **29**, 839TP; 1341TP.
Oster, G. (1950) *Compt. Rend.*, **232**, 1708.
Peters, L. (1960) *J. Text. Inst.*, **51**, T1290.
Peters, L. and Speakman, J. B. (1949) *J. Soc. Dyers and Colourists*, **65**, 63.
Peters, R. H. (1945) *J. Soc. Dyers and Colourists*, **61**, 95.
Peters, R. H. and O'Brian, C. D. (1953) *J. Soc. Dyers and Colourists*, **69**, 435.
Puffr, R. and Sebenda, J. (1967) *J. Polymer Sci.* (C), **16**, 79.
Remington, W. R. and Gladding, E. K. (1950) *J. Amer. Chem. Soc.*, **72**, 2553.
Scatchard, G. (1949) *Ann. N.Y. Acad. Sci.*, **57**, 660.
Schwarz, J. C. P. (1961) *Physical methods in Organic Chemistry*, Oliver and Boyd, London.
Steiner, R. F. and Beers, R. F. (1958) *Science*, **127**, 335.
Steiner, R. F. and Beers, R. F. (1959) *Arch. Biochem. Biophysics.*, **81**, 75.
Steinhardt, J. and Beychok, S. (1964) in *The Proteins* 2nd edition, ed. H. Neurath, Academic Press, London.
Steinhardt, J. and Harris, M. (1940) *J. Res. Nat. Bur. Stds.*, **24**, 335.
Steinhardt, J. and Reynolds, J. A. (1969) *Multiple Equilibria in Proteins*, Academic Press, London.
Stryer, L. (1965) *J. Mol. Biol.*, **13**, 482.
Vickerstaff, T. (1954) *The Physical Chemistry of Dyeing* Oliver and Boyd, London.
Watt, I. C. and Leeder, J. D. (1968) *J. Text. Inst.*, **59**, T353.
White, M. A. (1968) Ph.D. Thesis, Leeds University.
Yamaoka, K. and Resnik, R. A. (1966) *J. Phys. Chem.*, **70**, 4051.
Zanker, V. (1952) *Z. Physik. Chem.*, **199**, 15.

Chapter 6

The adsorption of dyes by cellulosic substrates

Although carbohydrates are as important in nature as proteins, the study of dye–carbohydrate interactions is virtually confined to textile substrates. Some interest has been shown in soluble carbohydrates as dye adsorbing species but the driving force of protein chemistry has no real parallel in this field. The question of the adsorption of dyes by cellulosic textile substrates is superficially complicated by the use of a variety of different kinds of dye, i.e. direct, vat, azoic and fibre-reactive dyes. These are applied by very different methods in each case and there is not the evident inter-relationship between them which is shown by acid and acid-milling dyes in the protein field. However the difference between these dyes in terms of the adsorption process itself are not qualitative. Certain physico-chemical aspects of other parts of the dyeing process are important however and special attention will be paid to aspects of vat dye reduction and oxidation in this section. Fibre-reactive dyes will be considered in a separate section.

I THE BINDING FORCES BETWEEN CELLULOSIC SUBSTRATES AND DYES

Cellulose consists of a polymer of 1:4 anhydroglucoside units.

Other carbohydrates vary in structure (as well as molecular weight) but all are polyhydroxylated substances which are strongly hydrophilic and generally water soluble except when strongly internally hydrogen bonded as in the case of cellulose itself. Many substrate materials of this kind, notably cellulose, contain carboxyl groups as a consequence of natural oxidation. These create an unfavourable electrical situation for anion adsorption. In any case the adsorption of an ion of either charge would determine the potential in a direction unfavourable to adsorption. High ionic strengths consequently characterize dye-substrate interactions of any notable order of magnitude. In the presence of water the compounds will be strongly hydrated and competition between water and dye molecules for hydrogen bonding sites would be expected to strongly disfavour dye-binding through such a mechanism. It would also seem at first sight that hydrophobic interactions are improbable in such a hydrophilic environment but some contribution from water structuring effects to dye binding is quite possible. Dispersion force interactions could make a significant contribution to adsorption. It is possible to draw some general conclusions from the clearly complex picture which is presented by substrates of this kind by considering the experimental evidence.

By tradition, cellulose (i.e. cotton) was dyed prior to the 1880's by natural and synthetic dyes on a mordant. Böttiger found, however, that if the molecular weight of simple acid dyes was built up by extending the conjugated system and that at the same time the molecules were preserved in a planar linear configuration the acid dyes so produced could be exhausted onto cotton from dye baths containing salt. The first of these *direct cotton dyes* was, of course, Congo Red. In modern terms it is clear that the simple acid dyes when adsorbed on cellulose will build up a negative electrical potential resisting further adsorption and that adsorption will only occur to the extent that the free energy gained by adsorption is not exceeded by the potential energy gained due to the charge repulsion effect. (This is similar to the situation on electrostatically saturated nylon.) The electrical effect can be offset by employing dyebaths of high ionic strength but clearly the lower the affinity of the dye the more electrolyte that must be used. What Böttinger found was that by using bis and tris azo dyes and keeping the charge down to a minimum, useful effects could be obtained. The important thing in coming to any conclusion about the binding mechanism is that the type of molecule which had to be used, i.e. highly conjugated and planar, provides the information that gross steric factors must be involved. The general conclusions from Böttiger's study are confirmed by the known importance of a significant degree of planarity in determining the affinity of leuco vat dyes and also the lack of influence of paraffinic weighting groups in dye molecules as compared with conjugated or dipolar substituents such as imido groups. The absence of any effect from paraffinic groups argues against hydrophobic interactions, as normally understood, playing any significant role in dye binding. Hydrogen bonding centres in the dye molecule do make a contribution to dye affinity. Much was made at one time of the correlation observed between the hydrogen bonding centre spacing in many direct cotton dyes and the unit cell repeat distance in cellulose as revealed by X-ray diffraction data (Robinson, 1954). However such a correlation is not only coincidence but also irrelevant. Not all satisfactory direct cotton dyes show the correlation and in any case this relates to crystalline regions into which the dye never penetrates. Allingham *et al.* (1954) have shown in monolayer studies that hydrogen bonding is not demonstrable between dyes and hydrated adsorbants such as carbohydrates. The marked need for linear or at least planar molecular configuration argues for close proximation between molecules and the polymer chain which is not essential for a hydrogen bonding mechanism. It does suggest that dipole interactions either of a permanent character between dipoles in the molecule or of the perturbation type (i.e. dispersion forces) are likely to be important.

Examination of some of the important groups in direct cotton dyes known to confer increased affinity reveals the presence of several dipolar groups e.g.

Benzoylamino groups

Ureido groups

Imido systems eg.

as well as the azo and other groups generally present as part of the chromogen.

The introduction of fibre-reactive dyes for cellulose has led to the use of a wide variety of heterocyclic substituents and it is clear that the more dipolar the character of the heterocycle the higher the affinity of the dye.

Evidence for the binding mechanism from adsorption studies is ambiguous since relatively little work has been done in single-phase systems. Hydrophobic bonding has been postulated by Iyer and Singh (1969) but the evidence is not definite. Also the hypothesis is, superficially at least, improbable. Their investigations were concerned with the interaction of Chlorazol Sky Blue FF and the substrates amylopectin and polyvinyl alcohol. The interaction was studied polarographically at two temperatures, 30°C and 40°C. The partial molal free energy of dye binding changed by only 0·5–1% over the temperature range. Nevertheless enthalpy and entropy calculations were carried out which purport to show the binding to be exothermic and with a positive entropy. Due to the paucity of experimental evidence the conclusions that hydrophobic bonding is involved seems tenuous. The work of Nishida, Akimoto and Uedira (1966) and of Carrol and Cheung (1962) was carried also at only two temperatures using Benzopurpurine 4B and Congo Red as dyes respectively on amylose and amylodextrin. The dye-binding was somewhat more temperature sensitive than in the case of the work of Iyer and Singh and over the range 18°C to 25°C the variation was 7%. The binding was endothermic and there was an entropy gain on binding. Nishida et al, (1966) suggest that the effect may be due to an increase in the number of available sites with temperature as a consequence of uncoiling of polymer chains in the substrate. No evidence is produced of temperature dependent saturation values for dye binding and Iyer and Singh produce definite evidence for no change of this kind. However randomization of polymer chains due to dye–substrate interaction would produce a considerable entropy gain and this may be the cause coupled with an associated breaking of hydrogen bonds to give endothermal adsorption. As will be considered in more detail at a later stage there is some other evidence for configurational changes in cellulosic polymers as a consequence of dye adsorption.

II   THE QUANTITATIVE TREATMENT OF DYEBATH EQUILIBRIA

Attention here is confined to the adsorption of anions by cellulosic fibres since this is the most frequently encountered situation. The discussion of this question in the literature is somewhat confused due to the introduction of several theories none of which until relatively recently provided an accurate description of experimental observations. Part of the problem arose from an inadequate understanding of the nature of a dye adsorption site and partly from the neglect of certain well established features of the theory of the electrical properties of surfaces.

The earliest attempt to rationalize the situation was due to Hanson, Neale and Stringfellow (1935) who treated the equilibrium as a three-phase situation involving an external and an internal aqueous phase and an adsorbing fibre phase. They assumed that the dye was distributed between the internal aqueous and fibre phases in accordance with a partition isotherm, i.e.

$$\frac{[D_f]}{[D_i]} = k \exp\left(-\Delta\mu^0/RT\right) \tag{6.1}$$

For a dye $Na_z D$, the value of $[D_i]$ may be obtained from the Donnan distribution equation since

$$\frac{[D_i]}{[D_s]} = \frac{[Na_s]^z}{[Na_i]^z} \tag{6.2}$$

Hence combining (6.1) and (6.2),

$$\frac{[D_f][Na_i]^z}{[D_s][Na_s]^z} = \exp\left(-\Delta\mu^0/RT\right) \tag{6.3}$$

If the sodium salt of the dye is the only electrolyte present so that other anions may be ignored, then neglecting any permanent changes on the cellulose, the value of $[Na_i]$ must be equal to the amount of sodium ion taken by the fibre expressed in g.ions/kg divided by the volume of the internal aqueous phase, V. Hence

$$\frac{[D_f][Na_f]^z}{V^z[D_s][Na_s]^z} = \exp\left(-\Delta\mu^0/RT\right) \tag{6.4}$$

$$\text{or } -\Delta\mu^0 = RT \ln \frac{[Na_f]^z[D_f]}{V^z} - RT \ln [Na_s]^z[D_s] \tag{6.5}$$

which is in experimentally accessible terms. The values of Na_f and D_f are expressed in g.ions/kg and of Na_s and D_s in g.ions/l.

In a later consideration of the problem, Crank (1947) took an apparently more sophisticated view of the situation again based on the three phases being present in the system. However Crank's derivation uses several simplifying assumptions which are in effect the same as those made explicitly by Hanson, Neale and Stringfellow.

In the first instance Crank assumes a Langmuir distribution of dye so that

$$[D_f] = \frac{f(-\Delta\mu^0) \cdot S \cdot [D_s]}{S + f(-\Delta\mu^0)} \tag{6.6}$$

and then because it seems reasonable to assume that saturation, $([D_f] = S)$, is never approached under normal experimental conditions

$$[D_f] = f(-\Delta\mu^0) \cdot [D_s] \tag{6.7}$$

Equation (6.7) is identical with a partition distribution equation. Crank then allows the partition isotherm to be modified to incorporate the Donnan distribution of sodium ions etc. and not surprisingly arrives at the earlier equation (6.6).

The consideration may be followed another way. If the affinity reflects the equilibrium between the internal aqueous phase and the fibre,

$$-(\mu_f^0 - \mu_i^0) = RT \ln a_f - RT \ln [D_i] \tag{6.8}$$

If the internal aqueous and external phases are regarded as ideal solutions then the standard states in the two must be the same, i.e. $\mu_i^0 = \mu_s^0$. Further the value of $[D_i]$ is given by the Donnan distribution equation (equation 6.2) so that

$$-\Delta\mu^0 = (\mu_f^0 - \mu_s^0) = RT \ln a_f - RT \ln [D_s] - RT \ln \frac{[Na_s]^z}{[Na_i]^z} \tag{6.9}$$

For a negative surface the value of $[Na_s]^z/[Na_i]^z$ is given by the Donnan theory,

$$\frac{[Na_s]^z}{[Na_i]^z} = \exp(-e\psi/kT) \tag{6.10}$$

so that

$$-\Delta\mu^0 = RT \ln a_f/[D_s] + e\psi/N \tag{6.11}$$

where ψ is the Donnan potential, $\bar{e}$ is the electronic charge and N is Avogadro's number.

The important feature of this analysis is the fact that it reveals the dependence of the affinity upon the surface potential of the adsorbing surface whatever is assumed about the value of the activity of the adsorbed dye. Equation (6.11) is essentially the same as that of Hanson, Neale and Stringfellow or of Crank but it is more general because of its fewer assumptions about the adsorbed dye ions and the relationship between $[Na_i]$ and $[D_f]$.

Since cellulose surfaces normally contain free carboxylic groups it is clear that the affinity value will be pH dependent except under low pH conditions of virtually complete back titration. Higher pH values designed to give complete carboxyl group ionization are likely to cause initial ionization of hydroxyl groups and a further increase in the electro negative potential. Thus at the high pH values of vat dye application for example, anions would be expected to exhibit lower affinity than under neutral conditions if all other factors were kept equal.

Another approach to the development of an equation describing the equilibrium is that known as the "diffuse adsorption model". In deriving values for internal pH etc. it is customary to treat all fixed charges as if they were dissolved in the diffuse double layer. The diffuse adsorption model treats all adsorbed species as if they were dissolved in the internal aqueous volume which corresponds with the diffuse double layer within the fibre. The activity of the adsorbed dye may be expressed as the ion activity product so that if the adsorbed species are measured in g.ions/kg, the internal volume is V and the dissolved species are expressed in g.ions/1 then, neglecting activity functions, for the dye $Na_z D$

$$-\Delta\mu^0 = RT \ln \frac{[Na_f]^z[D_f]}{V^{z+1}} - RT \ln [Na_s]^z[D_s] \qquad (6.12)$$

This equation is identical in form with that of Hanson, Neale and Stringfellow (6.6) and Crank (6.7). It is in essence based on the same kinds of concept although the assumptions made are apparently not the same.

The equations do broadly correspond to practice. If we take a pure dye $Na_z D$ then $[Na_f] = z[D_f]$ and $[Na_s] = z[D_s]$ so that equation (6.4) becomes

$$\frac{[D_f]}{[D_s]} = \left[V^z \exp(-\Delta\mu^0/RT) \right]^{1/(z+1)} \qquad (6.13)$$

which is a partition isotherm equation. This may be observed in very low liquor ratio conditions, e.g. in padding or in paper chromatography, when salt additions are not made. Under more normal dyeing conditions, i.e. longer liquor ratio conditions, this behaviour has been reported by Willis, Warwicker, Urquhart and Standing (1945) and also by Vickerstaff (1954).

Normally in the presence of salt, dyes on cellulose are observed to be adsorbed in accordance with a Freundlich isotherm. This is a statement which requires qualification because it normally means that the log/log relationship between $[D_f]$ and $[D_s]$ is linear. Due to the great insensitivity of such relationships to curvature the applicability of the Freundlich isotherm may be less general than supposed. However it is possible to reduce equation (4) to a Freundlich equation given certain assumptions as follows.

In a dyebath containing dye $Na_z D$ and sodium chloride, then within the fibre electrical neutrality exists and neglecting carboxyl groups in the cellulose,

$$Na_f^+ + H_f^+ = zD_f^- + Cl_f^- \qquad (6.14)$$

According to the Donnan distribution equations,

$$[H_f^+] = [Na_f^+][H_s^+]/[Na_s^+]$$
$$[Cl_f^-] = [Na_s^+][Cl_s^-]/[Na_f^+] \qquad (6.15)$$

so that if I is the ionic strength

$$\frac{[Na_f^+]I}{[Na_s^+]} = zD_f^- + \frac{[Na_s^+][Cl_s^-]}{[Na_f^+]} \qquad (6.16)$$

If the concentration of salt in the dyebath greatly exceeds that of the dye $[Na_s^+] = [Cl_s^-] = I$ so that

$$[Na_f^+] = \tfrac{1}{2}(zD_f^- \pm \sqrt{z^2 D_f^2 + 4I^2})$$ (6.17)

Under such high salt conditions exhaustion will normally be high so that $zD_f \gg I$.

Consequently

$$[Na_f] = zD_f$$ (6.18)

Substituting (6.18) in equation (6.12) gives

$$\frac{-\Delta\mu^0}{RT} = \frac{\ln zD_f^{z+1}}{V^z I^z [D_s]}$$ (6.19)

or $$D_f = k D_s^{1/z+1} \text{ where}$$ (6.20)

$$k = \frac{(VI)^z}{z} \exp \frac{(-\Delta\mu^0)}{RT}$$ (6.21)

The degree of correspondence between the predicted equations and the Freundlich isotherm cannot be regarded as good and may only be made to seem so by log/log plotting etc.

It is legitimate to query whether the equations employed to calculate affinities in dye–protein interactions would not provide a perfectly satisfactory alternative to those so far discussed. The use of such equations requires a knowledge of the concentration of available sites and unlike the case with proteins in which some chemically identifiable group is present at a concentration corresponding with maximum dye uptake, cellulose provides no obvious clue about saturation value except by direct measurement. This is unfortunately not easy because the saturation value is certainly high and in direct adsorption studies achieved only with highly concentrated and possibly highly aggregated solutions. The measurement of the saturation value of cellulosic fibres for dyes has something in common with the problem of measuring the saturation value of nylon with C.I. Acid Red 88. In the case of cellulose the problem is one of surface aggregation or multilayer adsorption. Neale and Stringfellow (1940) made a direct assult on the problem using Chorazol Sky Blue FF. From their data they calculated an apparent surface area of $1.5 \times 10^6 \times cm^2/g$ which was so far larger than data existing at that time that it was the general conclusion that the dye was aggregated in the adsorbed state. It is interesting to note that later work by Jeffries and by Iyer et al. (1968), which will be discussed, very largely supports the results of Neale and Stringfellow (1940).

The alternative and standard approach is to construct a reciprocal plot of dye on fibre against dye in solution to give an intercept equal to the reciprocal saturation value at $1/[D_s] \to 0$. This approach is complicated in the case of cellulose because of the Donnan potential effect. However if the salt concentration is moderately high the effect is supressed and the concentration difference between internal and external solutions is not significant. Rectilinear reciprocal plots were obtained using Benzopurpurine 4B by Hanson and Neale (1934) and by Meitner (quoted by Vickerstaff (1954)) with Durazol Red 2B. Differences were observed with different forms of cellulose but also with different dyes. Values fell between 0·02 and 0·10 moles/kg over the whole range of cellulosic substrates. However the sparsity of the data precludes speculation as to the causes of variation. The existence of variation itself caused Vickerstaff to conclude that the concentration term used in the diffuse adsorption and related models was more satisfactory an expression of adsorbed dye activity.

The use of the ionic activity product to describe adsorbed dye activity requires a value to be attributed to the V term (cf. equation (6.12)) unless the value of $[Na_f]$ is determined analytically. If the value of V is assumed to be constant the term V^z may be neglected but if the value of $[Na_f]$ is not to be measured by chemical analysis, it must be calculated as $[Na_f]/V$ by use of the Donnan equations and for this to be done V must be known. Analytical methods of determining $[Na_f]$ can be used. Farrar and Neale (1952) carried out such a study and demonstrated that at low ionic strengths the Donnan equations were not accurate. This is not surprising in view of earlier discussion of the theory of the double layer (Chapter 2).

Such analysis is difficult in any case and the problem of producing isothermal data by such a procedure is daunting. In any case the Donnan equations are well established in so many fields that the calculation of $[Na_f]/V$ from accessible data is obviously to be preferred. Hanson, Neale and Stringfellow (1935) took the view that the volume V corresponded with the saturation regain of the cellulose and employed a value of 0·22 1/kg for cotton. This corresponded to the much earlier value of Urquhart and Williams (1924). Vickerstaff comments (1954) that this approach is unsatisfactory because "there is no reason to suppose that the surface available for moisture adsorption will necessarily be the same as that for dye adsorption". The preferred approach was one based on actual dye adsorption behaviour. Hanson, Neale and Stringfellow (1935) examined the adsorption of Chlorazol Sky Blue FF on a number of cellulosic materials at different salt concentrations. The materials had different carboxyl group contents and hence Donnan effects on the equilibrium. However taking the value of $[D_f]$ on cotton cellulose as a reference it was noticed that the relative dye uptake by the different celluloses was salt sensitive below about 3% sodium chloride concentration. The data of Table 6.1 are calculated from their data.

Vickerstaff argues that providing sufficient salt is present to supress the surface potential difference, the relative volume terms may be calculated. This

TABLE 6.1.   Relative dye uptake values by different celluloses.

Relative value of $[D_f]$

% Sodium Chloride in solution	Cotton	Mercerized Cotton	Hydro-cellulose	Oxy-cellulose	cello-phane	Viscose rayon	Cupram-monium rayon
0·25	1·00	—	—	0·17	0·50	—	—
0·50	1·00	1·80	—	0·20	0·80	—	1·90
1·00	1·00	1·65	0·77	0·18	1·29	1·18	2·35
2·00	1·00	—	—	0·29	1·75	—	2·50
5·00	1·00	1·53	0·78	0·43	2·29	2·12	2·71
10·00	1·00	—	0·85	0·58	2·31	—	2·64
20·00	1·00	1·56	—	0·71	2·32	—	2·63
35·00	1·00	1·62	0·80	—	2·29	—	—

led to values of 0·45 1/kg for viscose rayon using 0·22 1/kg for cotton as a basis.

Marshall and Peters (1947) employed another method for obtaining values of the V term. Isotherms were obtained, Chrysophenine G on cotton, viscose rayon and cuprammonium rayon and the values of $[Na_f]/V$ calculated using arbitrary values of V so as to give a rectilinear relationship of unit slope between the ion activity product terms for the two phases, cf. equation (6.5).

Vickerstaff comments that the ionic strengths employed (0·01 N–02) were not sufficiently high to offset the surface potential differences between the fibres and also that the log/log relationship is very insensitive to changes in V. The values obtained are shown in Table 6.2.

TABLE 6.2.

Fibre	Calculated volume term
Cotton	0·30 1/kg
Mercerized cotton	0·50 1/kg
Viscose rayon	0·45 1/kg
Cellophane	0·45 1/kg
Cuprammonium rayon	0·65 1/kg

Using these values Marshall and Peters calculated affinities for some fourteen dyes on different fibres and obtained for any given dye fairly constant affinity values. They came to the conclusion that the binding energy of a dye for cellulose was characteristic of the interaction only at a molecular level and consequently variables such as orientation or physical differences which mark the differences between different forms of cellulose were irrelevant. This view was supported by the constancy of the activation energy of adsorption of six of the dyes examined on different substrates.

Attractive though this viewpoint may be, there are a number of experimental observations which do not fit with it and consequently it is not sufficient to leave matters at the stage reached by the work of Marshall and Peters. In the first instance the varying effect of carboxylic groups in the different forms of cellulose is neglected in their work although at heavy depths of shade the concentration of bound dye ions may exceed that of bound carboxyl groups.

Most of the experiments were not carried out at sufficiently heavy depths for this condition to be assured however. The second factor which has been neglected until recently is that even if the concentration of surface charges of the surface potential effects are neglected or minimized the value of V will depend upon the ionic strength because this determines the thickness of the diffuse double layer. Consequently the value of V must be regarded as variable rather than constant in any calculations. In addition there have been results obtained which tend to cast some doubt on the validity of the concept of a constant affinity for cellulose.

Graham and Fromm (1947) calculated heat of dyeing values from the data of Willis et al. (1945) and from their own experiments and found that the heat of dyeing decreased as the dye concentration on the fibre increased. This could indicate surface heterogeneity as well as other possible factors but in any case it is an observation which goes against the conclusions of Marshall and Peters. In another study Nishida (1951) found that the volume term value required to rationalize data from one direct cotton dye was far lower than that required for another. In all these studies so far discussed the problem of the activity of the dye in solution remains a factor creating doubt in developing any theory particularly where heat of dyeing data and heavy shade applications are concerned. However recent studies have enabled the outstanding problems to be settled fairly satisfactorily including the vexed question of the volume term.

In an exhaustive study of three direct dyes on cotton, viscose rayon and cuprammonium rayon, Daruwalla and D'Silva (1963) produced data from which heat of dyeing values as a function of $[D_f]$ could be calculated. The method used has been described (Chapter 2) and a typical relationship is shown in Fig. 6.1.

The results suggest that there is present a small number of very active sites but there may be impurities in the cellulose. Their possible presence certainly requires further study. The dominant features of the relationship are, however, a region of constant enthalpy terminated by a sharp break. This was observed in all cases. The constant enthalpy values were not the same for all forms of cellulose although not markedly different. The breaks in the enthalpy/concentration plot occurred at quite moderate shade depths, e.g. in the above example at 1–2% depth of shade. Daruwalla and D'Silva suggest that the breaks mark surface saturation so that at higher values of $[D_f]$ the characteristic interaction is multilayer rather than dye–cellulose interactions.

Assuming molecular surface area values for the dyes Daruwalla and D'Silva calculated adsorbing surface areas for the different forms of cellulose in the cases of the three dyes. Not surprisingly the dyes gave different area values but the relative areas for the three celluloses were found to be independent of the dye. This is shown in Table 6.3.

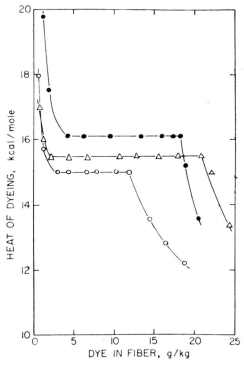

Fig. 6.1.   Typical variation of ΔH with $[D_f]$ with Chrysophenine G (reproduced with permission
from Daruwalla and D'Silva, 1963).
● Viscose rayon; ⊖ cotton; △ cuprammonium rayon.

TABLE 6.3.   Calculated surface area values for different cellulose (Daruwalla and D'Silva, 1963).

	Surface area values (m^2/g)			
	Chlorazol Sky Blue FF	Congo Red	Chrysophenine G	Relative area value
Cotton	22	31	39	100
Viscose Rayon	32	45	58	150
Cuprammonium Rayon	39	53	66	170

The surface area values are all far greater than values determined for dry
cellulose but of the same order of magnitude as obtained from water adsorption values (Jeffries, 1960). The relative area values are comparable with those
attributed to Meitner by Vickerstaff (1954).

Further information regarding saturation of the cellulose surface was
obtained in a further study by Iyer and co-workers (1964). In a detailed study
of the isotherm for Chlorazol Sky Blue FF on viscose rayon isotherms of the
type shown in Fig. 6.2 were obtained.

Isotherms of the same kind are characteristic of surfactants adsorbed on
textiles or polar adsorbants (Evans, 1958; Tamamushi and Tamamatu,
1959). The break in the isotherm in such cases corresponds to the critical

micelle concentration (CMC) of the surfactant in solution. Arguing by analogy Iyer *et al.* (1964) suggested that any consideration of the Chlorazol Sky
Blue FF isotherm must be confined to the lower concentration region. Further more it was argued that the plateau region of the isotherm represented a
saturation of the adsorbing cellulose surface so that a Langmuir equation
should be employed to describe the equilibrium up to that point. The equilibrium in question is an internal one between dye on the fibre surface and in

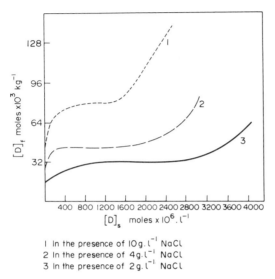

Fig. 6.2.   Adsorption isotherms on cellulose, Chlorazol Sky Blue FF (Iyer *et al.*, 1964).

the internal volume. In contrast with earlier investigators Iyer *et al.* regarded
the internal volume as being a variable equal to the area available for
adsorption of water times the thickness of the double layer given by Verwey
and Overbeek (1948).

$$\frac{1}{2e}\left[\frac{2\pi[I]\,N\times10^{-3}}{\varepsilon kT}\right]^{-1/2} \tag{6.22}$$

where e is the electronic charge, ε is the dielectric constant, N is Avogadro's
number, k is the Boltzmann constant, T is the absolute temperature and I is
the ionic strength. The available area was calculated using the B.E.T. isotherm equation from the data of Jeffries (1960).

   Using the V term thus calculated the concentration of dye in the internal
phase, $[D_i]$ was calculated using the Donnan equations. The internal Langmuir equation,

$$\frac{1}{[D_f]}=\frac{K}{S}\cdot\frac{1}{[D_i]}+\frac{1}{S} \tag{6.23}$$

in which the Langmuir adsorption constant, K is related to the affinity by

$$-\Delta\mu^0 = -RT \ln K \qquad (6.24)$$

was used to calculate the affinity and saturation values.

In order to employ equation (6.23) the assumption was made that $[D_f]$ was equal to the *total* dye adsorbed in the same manner as Gilbert and Rideal in dealing with proteins. The error involved is not significant. It was found that plots of $[D_f]^{-1}$ against $[D_i]^{-1}$ were rectilinear in all cases in contrast to the usual Freundlich relationship obtained when the external dye concentration is used. The affinity values and saturation values were similar to those obtained by Daruwalla and D'Silva (1963). Calculations were made of data in which the role of carboxyl groups in the substrate or the variation of V with μ were ignored (i.e. using $V = 0.45$ 1/kg). The resulting discrepancies were small although absolute values were changed. From the saturation and surface area values, limiting co-areas (i.e. minimum occupied surface areas per molecule) were calculated. These were found to be dependent upon the ionic strength.

Although this investigation did not provide any justification for the use of a variable V term, further studies by Iyer and Jayaram (1970) in which Donnan potentials of viscose films were measured showed the dependence of the potential and hence the thickness of the double layer on $I^{1/2}$. This was shown to be true for a wide range of depths of shade and ionic strengths. Although the correction is not large it is clearly better to employ the correct term for the calculation of V.

Further studies (Iyer *et al.*, 1968) were concerned with the role of the cation in the dyebath. The adsorption of Chlorazol Sky Blue FF by cotton fibres from dilute solutions of the alkali metal chlorides was studied. It was found that while the affinity of the dye anion was unaffected by the size of the cation, the saturation value (and hence the limiting co-area) were markedly affected. This was explained on the basis that the water structure breaking effect of the cations was known to increase with ionic size. This would of course provide an alternative explanation.

The variation of the limiting co-areas of adsorbed dye molecules with ionic strength is an important observation and has received special attention from Iyer and Baddi (1968). From enthalpy calculations relating to the adsorption of Chlorazol Sky Blue FF by cotton, viscose rayon and cuprammonium rayon and also calculated limiting co-area values, a theoretical treatment was developed which confirms much of what has been said about adsorption sites and interestingly relates with the consideration of dye adsorption by polyamides. The relationship between enthalpy and limiting co-area is shown in Fig. 6.3.

It can be seen that all three cellulosic substrates behave identically and that the enthalpy falls with limiting co-area as the ionic strength decreases.

The value of the affinity was observed to vary negligibly with the limiting co-area and consequently it was concluded that the enthalpy variation must be compensated by an entropy change in each case. Iyer and Baddi suggest that the role of electrolyte is, in addition to supressing the potential developed at the charged cellulose surface to a limited extent, a suppression

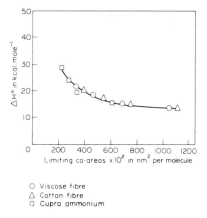

○ Viscose fibre
△ Cotton fibre
□ Cupra ammonium

Fig. 6.3.   The relationship between enathalpy of adsorption and the limiting co-area (Iyer and Baddi, 1968).

of coulombic dye–dye repulsions within the surface. This is clearly similar to the potential energy effect discussed in relation to polyamides. Since the surface potential effect should have no enthalpic contribution, its reduction should lead to an increase in the effective enthalpy of adsorption. The closer arrangement of adsorbed molecules (smaller limiting co-areas) must be accompanied by a relative loss of entropy of the same order of magnitude. This situation has been observed elsewhere (Barrer and Wasilewski, 1961) and has become a generally expected one for adsorption on charged surfaces (Devanathan, 1962).

If the saturation values of cellulosic substrates are a function of the total charge density on the adsorbing surface, it is clear that where anionic dyes are applied under alkaline conditions such as to cause ionization of hydroxyl groups, there will be a considerable need for high ionic strength dyebaths to bring about good dyebath exhaustion. Such conditions are met during the naptholation stage of azoic dyeing where the naphthol is applied as the ionized sodium salt from alkaline solution. They are also met in vat dyeing where the ionized leuco salt is applied from high pH dyebaths. If in determining the affinity of such dyes for cellulose under alkaline conditions an uncorrected affinity is calculated (i.e. the effect of surface potential is ignored) then the apparent affinity will be low. From equation (6.11), the apparent affinity is

$$(-\Delta\mu^0)_{apparent} = -\Delta\mu^0 + \frac{e\psi}{N} + RT \ln \frac{a_f}{D_s} \tag{6.25}$$

Uncorrected leuco vat dye affinities have been calculated by Fowler, Michie and Vickerstaff (1951) and shown to be low as compared with direct dyes. Dimethoxy dibenzanthrone (Caledon Jade Green XN) was found to have an affinity of 5·33 kcal/mole at 40°C comparable with that determined for Chrysophenine G. The latter possesses low affinity as a direct cotton dye. The sulphate ester of the leuco form of Caledon Jade Green XN (i.e. the corresponding Indigosol dye) exhibits very high substantivity for cellulose. Its affinity has never been measured but it is quite clearly much higher than that of Chrysophenine G under neutral application conditions. The marked difference in behaviour between the two forms is indicative of the surface potential effect. It is to be expected in view of Iyer's results that equation (6.25) will underestimate the effect of surface ionization. It has been seen how the greater the surface potential energy or charge repulsion effect in the surface the lower the saturation value. If the fibre surface is ionized then the effect will be similar to that of dyeing from a lower ionic strength solution. If the true (corrected) affinity is independent of the ionization then for a given value of $[D_s]$ the value of a_f in equation (6.25) must remain the same. Since the dye is adsorbed in accordance with a Langmuirean model, the value of a_f is given by

$$a_f = \frac{[D_f]}{S - [D_f]} \quad \text{(cf. equation 2.6)}$$

so that for a constant value of a_f, $[D_f]$ must fall with S, the saturation value. The practical effect is consequently a lower apparent affinity value when calculated on the basis of the Hanson, Neale and Stringfellow equation (i.e. $a_f = [D_f]/V$).

Some further evidence for the intrinsic lowering of affinity values due to the presence of negatively charged groups can be gleaned from an examination of the results of Aspland and Bird (1961) on the adsorption of non-ionic dyes by cellulose. Such dyes were found to give partition isotherms on cellulose as on other substrates. The molecular weights of the compounds used were necessarily lower than those of direct dyes due to the need for adequate aqueous solubility. However thirteen compounds were studied and it was possible to estimate a number of substituent effects which were found to be additive. Using this data it is possible to build up a non-ionic "analogue" of Chrysophenine G as shown below.

Chrysophenine G $-\Delta\mu^0 = 3\cdot80$ k cal/mole at 60°C (Vickerstaff, 1954)

Non-ionic analogue   Predicted $-\Delta\mu^0 = 4\cdot46$ k cal/mole at 60°C

$$0\cdot31 + 1\cdot74 + 0\cdot36 + 1\cdot74 + 0\cdot31$$

A value of $0\cdot36$ kcal/mole is ascribed to the stilbenyl group by assuming it to be equivalent to an azo group.

Although such a calculation is necessarily very approximate the difference is sufficiently large in comparing the dye with its analogue to support the hypothesis that the charge centres in the dye lower the affinity by charge repulsion effects.

III   COMPLEX DYEING SYSTEMS FOR CELLULOSIC SUBSTRATES

The problem of achieving fastness to wet treatments in cellulosic fibre dyeing has given rise to several forms of dye which are applied in the course of a fairly complex sequence of operations of which dye adsorption is but one. The adsorption stage in the overwhelming majority of cases is not different from that already discussed but the other aspects of the application sequence is in three cases, i.e. vat, leuco ester and fibre-reactive dyes, of significant physico-chemical interest. Of these the two relating to vat and ester dyes will be considered here. Fibre-reactive dyes merit a special treatment. Nevertheless all three have one feature in common and that is the incorporation into the dyeing process of chemical treatments to modify the dye.

A   Leuco ester dyes

These are the sulphate esters of the leuco quinonoid vat dyes e.g.

C.I. Solubilized Vat Green I (Soledon Jade Green X)

They are applied by padding or exhaustion methods and the vat pigment is regenerated by an acid oxidation treatment.

Early theories of the mechanism of the oxidation of these compounds was based on a hypothetical saponification of the ester,

(a) $NaSO_3O:D\cdot OSO_3Na + 2H_2O \rightleftharpoons HO\cdot D\cdot OH + 2NaHSO_4$

(b) $HO\cdot D\cdot OH + O \rightarrow O = D = O + H_2O$

This concept was first rejected by Bader (1938) on the grounds that if the equilibrium of reaction (a) played a significant role in such a scheme then the acid leuco dye ($OH\cdot D\cdot OH$) would precipitate from leuco vat ester solutions in view of its very low solubility, and that simple treatment of the acid leuco dye with sulphuric acid would yield the ester. In fact leuco vat ester solutions are very stable and rigorous conditions are necessary to form them (Bader, 1924).

$$HO\cdot D\cdot OH$$
$$\Big|2HSO_3Cl$$
$$\downarrow 2C_5H_5N$$
$$[SO_3\cdot O\cdot D\cdot O\cdot SO_3]^= (C_5H_5N^+)_2$$
$$\downarrow 2NaOH$$
$$Na_2^+[SO_3\cdot O\cdot D\cdot O\cdot SO_3]^= + 2C_5H_5N + 2H_2O$$

The mechanism of oxidation has been elucidated only relatively recently by Johnson and several co-workers (1955, 1957, 1958, 1960). It was found to be a very complex reaction which could take a number of routes according to concentration conditions. Initial studies using hydrogen peroxide showed that the rate of oxidation under acid conditions occurred more rapidly than the sulphate groups could be removed by hydrolysis alone. Moreover hydrogen peroxide which has oxidizing power under acid and alkaline conditions was found to work only under the former. Ferrous salts which produce both hydroxyl and perhydroxyl radicals in interaction with hydrogen peroxide were found to accelerate oxidation very considerably leading to the conclusion that such radicals play an important role in a direct oxidative mechanism involving the dye ester. Detailed studies showed that in peroxide oxidation a vital step was played by an irreversible partial hydrolysis to form the mono ester of the acid leuco dye.

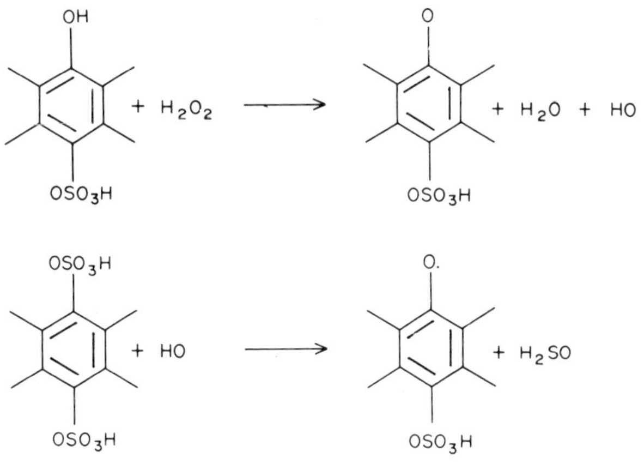

Only one of the several modes of action of hydroxyl radical are shown here. The radical can behave in various ways depending upon the concentration of hydrogen peroxide. However the result of the complex possibilities is the direct oxidation of the ester. The catalytic effect of the monoester was demonstrated by its synthesis and addition to oxidation reaction systems. This observation also serves to demonstrate that the formation of the mono-ester is non-reversible.

Normally leuco vat esters are oxidized by acid nitrate solutions and this introduces further complications to a description of the mechanism of oxidation. With this oxidizing agent, oxygen also plays a direct role if present.

In the complete absence of oxygen, leuco vat esters were observed to be oxidized by nitrous acid solutions at the same rate as the ester groups were hydrolysed. In the presence of oxygen reaction rates were much faster. Oxygen dissolved in the reagent solutions promoted very fast but limited reaction consuming the oxygen. Addition of oxygen to reacting solutions promoted greatly increased reaction rates. Aspland and Johnson demonstrated very neatly the factors operating in the system by using varying experimental conditions and established the following as the oxygen catalyzed sequence.

(a) $HO \cdot SO_3 \cdot D \cdot SO_3OH + 2H_2O \rightarrow HO \cdot D \cdot OH + 2H_2SO_4$
(b) $HO \cdot D \cdot OH + O_2 \rightarrow O = D = O + H_2O_2$

Reactions (a) and (b) in the presence of a limited amount of oxygen represents a consumption of that available. However the oxidation can produce further oxidation through the reaction of the hydrogen peroxide with nitric acid.

(c) $H_2O_2 + HNO_2 \rightarrow HOONO + H_2O$
(d) $HOONO \rightarrow HO \cdot + NO_2$

The hydroxyl radical can cause a further oxidation of ester directly as already discussed. It can also recombine with nitrogen peroxide.

(e) $HO + NO_2 \rightarrow HNO_3$

Thus each molecule of oxygen can cause the oxidation of more than one leuco ester molecule. However the chain of reaction is limited by reaction (e) which functions as a chain stopper.

Reactions (e) and (d) are confirmed by other studies (Halfpenny and Robinson, 1952) and the presence of hydroxyl radicals in the system under consideration was elegantly demonstrated by the catalysis of the polymerization of acrylonitrile. This is known to be catalysed by hydroxyl radicals but not by nitrogen oxides or acids. The polymerization was catalysed by peroxide/nitrous acid systems with the consumption of nitrous acid. The extent of propagation of the reaction suggested by the above scheme is however insufficient to account for the observed reaction rates in the presence of oxygen. Aspland and Johnson proposed another reaction competing with reaction (e)

(f) $NO_2 + HNO_2 \rightarrow HOONO + NO$

Such a reaction would certainly provide the necessary propagation effect. Evidence for the hypothesis is not easy to obtain. Addition of nitrogen dioxide to a solution of a leuco vat ester and nitrous acid should accelerate oxidation. This was found to be the case. Unfortunately pure nitrogen dioxide cannot be used because of its tendency to dimerize. This yields nitrosonium ions (NO^+) in contact with water which react to give nitrous acid. The expected consumption of nitrous acid through reaction (f) cannot be observed. The contrary is in fact to be expected and was found. The circumstantial evidence is consequently strong.

The mechanistic basis for industrial practice is now clear. A large excess of sodium nitrate is employed which inevitably favours the reaction propagation reaction, (f) rather than the chain-stopping reaction (e). In addition oxygen is freely available both dissolved in the reagent solutions and through deliberate exposure of the fabric after acidification. Aspland and Johnson recommend the addition of hydrogen peroxide as a consequence of their observations. It is not known whether practical dyers have shown any interest in this suggestion.

B   Vat dyes

These are quinonoid compounds of very low solubility in water, applied by reduction of the dispersed pigment in water to give a solution of the ionized leuco compound. This is subsequently oxidized to the original pigment and the material washed off. Although they are of declining importance as dyes in Europe, they remain of importance in countries where hypochlorite bleach is used extensively in laundering and also in the USA where very long high speed production runs (10^6 metres) demand a very high level of reliability in use during dyeing. The latter property is a most notable advantage of vat dyes and suggests that they are by no means technically obsolete. The process of solubilization by reduction followed by oxidation on the fabric is much cheaper than that involving esterification as with the leuco vat ester dyes although the final dyed product is identical. Consequently vat dyes and vat dyeing are of far greater technical importance than the leuco ester dyes. Since the dyeing process commences with a pigment the actual dyeing process is one of five steps.

* dispersion of the pigment
* reduction to the leuco compound
* adsorption
* oxidation
* clearing or soaping

Each of the five steps will be considered in turn.

(1) *Dispersion of vat pigments*

This stage in the dyeing process has become of importance only in the relatively recent period. Vat dyes have, through the use of indigo, a venerable history and for centuries the vat pigment was simply stirred up in water with appropriate reducing compounds. Treatment was continued until reduction was complete, i.e. complete solution effected. Within broad limits the dispersion properties of the pigment was not of great importance. Synthetic vat dyes became of increasing importance during the first three decades of this century without significantly affecting the situation. Manufactured vat dyes required a certain amount of grinding to give a free flowing powder but there could not be said to be any real concern with rapidity of reduction or uniformity of reduction rate through the pigment sample beyond a certain obvious level. Two factors changed this situation, the development of package dyeing and (of greater importance) padding methods of application to piece goods, e.g. pad-dye jig development and continuous dyeing. The important feature uniting these somewhat disparate dyeing methods is the application of the vat pigment dispersion directly to the material, followed by its reduction (and dyeing) in situ. In package dyeing, a fine uniform dispersion of vat pigment is circulated through the package of yarns and controllably cracked out to deposit the pigment on and in the yarns. In the padding processes the fabric is impregnated with a dispersion of the pigment in a padding mangle and usually dried.

Clearly in processes of this kind uniformity of the pigment dispersion is of great importance in order to obtain a satisfactory distribution of dye pigment through the material. This requirement also sets an upper limit on particle size. In addition where continuous processing is involved the rate of reduction will affect the productivity of the process, a small particle size will be desirable so as to present the maximum surface area in the heterogeneous reduction reaction. Finally the stability of the dispersion will be of importance in slightly differing ways according to the details of the application sequence. The problem of the colour chemist has been to concern himself with these factors with a view to providing the dyer with a pigment powder which simply has to be added to water and stirred lightly. All the milling techniques and other methods used in dispersing pigments in paint media are too complex for the conditions of textile colouration and the "instant" approach to dispersion is required.

Three stages are of importance in the dispersion process;
wetting: The air at the powder surface must be rapidly and virtually completely replaced by liquid so as to avoid flotation etc.

deagglomeration: crude aggregates of pigment powder must be readily broken up to give the required degree of dispersion.

stabilization: the conditions must be such that the normal forces of attraction between dispersed particles in an unstabilized dispersion cannot operate, i.e. ions must be adsorbed to give charge repulsion between particles or molecules taken up from solution which modify the surface attraction forces.

The stages may occur very rapidly and virtually simultaneously. Clearly the third stage cannot lag far behind the first two stages and the second is at least in part a consequence of the first. For very rapid easy dispersion hydrophobic particles such as vat pigments must be treated with dispersing or wetting agents and agglomeration kept at an absolute minimum in manufacture. Stability is achieved mainly by adsorption of appropriate surfactants on pigment particles but also charge effects due to ion adsorption may play a part. Additionally it is clear that all other factors being equal, the smaller the particle size the more stable will be the dispersion. It can be technically embarrassing to have too stable a dispersion however.

In the yarn pigmentation dyeing process the dispersion must be controllably broken down. The addition of electrolyte and heat are obvious means for achieving this due to their reduction of coulombic charge repulsion effects or modification of the state of solution of surfactants. However dispersions can be made which are so stable that pigmentation of the yarns is impossible or necessitate excessive electrolyte additions. In padding processes the problems due to excessive dispersion stability arise when the impregnated fabric is dried. Possessing no affinity for the material pigment particles tend to move with water migrating to the fabric surface in the drying process.

A really satisfactory dispersion for padding application will tend to break down rapidly in drying so that the pigment particles remain still while water migration proceeds.

Thus a satisfactory vat pigment dispersion represents a compromise solution between opposing needs. In general a particle size which is reasonably uniform and of the order of 2μ is found to be satisfactory. Dispersing agents and preparative techniques are closely guarded industrial "secrets" and very little has been published in this area other than commercial or techno-commercial information.

(2)   Reduction to the leuco compound

Although most investigators have been concerned with the reduction/ oxidation equilibrium of vat dyes, it is in fact the rate of reduction of vat pigment dispersion which is of the greater technical interest. The study of this parameter is made difficult by the heterogeneous nature of the reduction and the interplay of the various factors which can effect the reduction rate, namely,

(a)   the chemical constitution of the vat pigment
(b)   the particle size
(c)   the crystal habit
(d)   the temperature
(e)   the concentration of reagents.

In addition, due to the reversible nature of the reduction process,

$$C=0+e\rightleftharpoons C-0^-$$

the reaction rate would be expected to be diffusion controlled, i.e. to depend upon the rate of removal of reduced leuco compound from the pigment surface.

If an idealized system is considered, i.e. the reduction of spherical particles of uniform diameter under fixed temperature and reagent concentration conditions, it is possible to treat the reduction as a peeling process or the removal of successive layers of molecules by reduction.

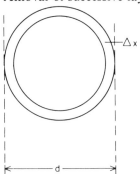

Consider a system of n particles of diameter d which contain m molecules per unit volume and in which a mono molecular layer has a thickness Δx.

At any time, t, the number of molecules available for reduction at the surface of the particle is

$$n\,m\left[\frac{\pi d_t^3}{6} - \frac{\Pi(d_t - 2\Delta x)^3}{6}\right] \tag{6.27}$$

Since Δx is very small, its higher powers in the expansion of (6.27) may be neglected so that the number of molecules available for reaction, i.e. the effective concentration, will be

$$n\,m\,\pi\Delta x d_t^2 = \phi d_t^2 \text{ where } \phi = nm\pi\Delta x \tag{6.28}$$

If the reversibility of the reduction reaction is ignored and it is assumed implicitly that all reduced molecules are instantaneously removed from the surface, the rate of reduction will be given by

$$\frac{-dw}{dt} = k_1\,\phi d_t^2 \tag{6.29}$$

in which $-dw/dt$ is the rate of loss of the weight of the unreduced vat dye particles, k_1 is the velocity constant. It is implicitly assumed in (6.29) that the concentrations of other reagents are kept constant. It is not possible to integrate this equation directly but since w and d are interrelated through the particle density σ it is possible to transform equation (6.29) to a soluble form

$$w_t = \frac{\sigma \cdot \Pi d_t^{\,3}}{6} \tag{6.30}$$

so that substituting for d_t^2 in equation (6.29)

$$\frac{-dw}{dt} = k_1\,\phi\left[\frac{6}{\sigma\Pi}\right]^{2/3} w_t^{\,2/3} \tag{6.31}$$

Also from equation (6.30)

$$\frac{dw}{dt} = \frac{\sigma\Pi d_t^{\,2}}{2} \tag{6.32}$$

Therefore, combining (6.32) and (6.29),

$$\frac{-dd}{dt} = \frac{-dw}{dt}\cdot\frac{dd}{dw} = k_1\sigma d_t^2\cdot\frac{2}{\sigma\pi d_t^2} = \frac{2k_1\,\phi}{\sigma\pi} \tag{6.33}$$

Thus the prediction based on the assumptions made is that the rate of change of weight of unreduced vat pigment will be proportional to $w_t^{2/3}$ and that the rate of change of particle diameter will be constant. The first prediction (equation 6.31) was found by Marshall and Peters (1952) to be incorrect. The rate of reduction was found to be greater than predicted and to conform the first order kinetics, i.e.

$$\frac{-dw}{dt} = k' w_t \tag{6.34}$$

In deriving equation (6.31) two broad assumptions are involved apart from those involving the geometry of the system. The first is that the only mechanism for the dissolution of particles is reduction to the leuco compound. The second is that diffusional factors are not involved. Of the latter there are two possibilities namely diffusion of the leuco compound away from the particle surface and diffusion of the reagents across the interfacial boundary into the solid phase. These two main assumptions may be considered in turn.

It has been known for many years that very small particles possess abnormally high solubility. For uncharged particles the relationship between radius and solubility is given by

$$RT \ln \frac{S}{S_0} = \frac{2\gamma \bar{V}}{r} \tag{6.35}$$

in which S, S_0 are the solubilities of particles of radius r and of infinite radius respectively. $\bar{V}$ is the molar volume and γ is the surface tension in the solid. Equation (6.35) is based on Kelvin's equation for vapour pressures near curved surfaces and assumes ideality in the system. Because a suspension of particles of abnormally high solubility is subject to particle size growth, i.e. less soluble particles grow at the expense of more soluble particles, equation (6.35) is not an easy one to verify in any case. However abnormal solubilities are well established, e.g. Barium sulphate particles of average size 0.1μ show as much as 80% excess solubility (Dundon and Mack, 1923). There are modifications to equation (6.35) such as that due to Knapp (1922) which allows for charges on the particles but at this level of particle size systems are inherently unstable and very difficult to study. Nevertheless for the purpose of the present discussion it is clear that particles will disappear more rapidly than would be expected if reduction alone was the cause.

The two diffusional factors potentially have opposite effects on the reduction rate. At the particle surface an equilibrium will exist,

$$\text{oxidized form} \underset{k_2}{\overset{k_1}{\rightleftharpoons}} \text{reduced form.}$$

The rate of reduction will be equal to

$$k_1 \left[\text{oxidized form}\right] - k_2 \left[\text{reduced form}\right]$$

and if [reduced form] is kept very small due to very rapid diffusion then the optimum rate will be observed. The value of k_1 [oxidized form] is a constant for a given set of conditions and the rate must fall as [reduced form] increases.

The flux away from the article surface is given by the Fickian equation,

$$\frac{ds}{dt} = -D \cdot A \cdot \nabla c = \frac{-\bar{D} \cdot A \cdot [\text{reduced form}]}{x} \tag{6.36}$$

in which $\bar{D}$ is the diffusion coefficient, A is the cross section of the diffusional process (i.e. the surface area of the spherical particle), x is a constant distance within the value of [reduced form] has fallen effectively to zero. The term [reduced form] gives the concentration of the leuco dye at the particle surface. Since the flux is a function of the particle surface area and since this is proportional to the square of the diameter, it must fall as the particles grow smaller. Combining this concept with equation (6.29) gives

$$\frac{-dw}{dt} = \frac{k_1 \phi d_t^2}{\alpha d_t^2} = \frac{k_1 \phi}{\alpha} \tag{6.37}$$

in which α is a combination of diffusional constants. The reaction is thus predicted to be of zero order and slower than predicted by the simple theory. Clearly the influence of this diffusional factor must be small, i.e. ds/dt is large enough to be insignificantly changed by surface area. Since the vat pigment surface is hydrophobic and the leuco compound is hydrophilic this seems likely to be the case.

The other diffusional factor is concerned with the rate of diffusion of reagents from the solution phase across the interface between the particles and the solution. This is related to the contact between the two phases or the interfacial energy. It would be expected as a consequence that the rate of reduction would be changed due to the presence of surfactants and this is found to be the case. Surfactants have been examined in vat pigment reduction on a technical basis because of their useful catalytic effect. The first to be examined were the quinone sulphonic acids particularly anthraquinone β-sulphonate. This was thought to act catalytically by direct participation in the reduction

Such a mechanism may well play some part but it is more probable that the surface active properties of this compound are mainly responsible for the effect. This certainly is true of polyethanoxy compounds such as Dispersol VL which are very effective (Rattee and Seltzer, 1950). In many cases vat pigment dispersions are sufficiently fine and stable for them to be treated as solutions obeying Beer's Law. In such cases vatting rates and catalytic effects are easily followed in a spectrophotometer fitted with a thermostatted cell system.

From what has been said it is clear that a marked effect on reduction rate due to crystal habit is to be expected. This would affect the surface energy of the crystals, the disposition of the molecules in the surface and the hydrophobicity of the surface. In their study Marshall and Peters (1952) found marked effects due to crystal habit which were more important than particle size variations. They also found that particle size effects were difficult to study because large particles are often composed of microcrystallites and may break down physically during reduction. The quantities of reducing agents are normally in excess in vatting and consequently vatting rates are not sensitive to their concentration. Chemical constitution was thought to have a major effect on reduction rate. This is not unexpected but it is a situation which is very difficult to rationalize because of the contribution of particle size and crystal habit. The effect of constitution could be isolated by following reduction rates in solution but this device would be limited in value because of the limited number of solvent soluble vat dyes and suitable solvents. The effect of temperature has been studied by Marshall and Peters and by Belenkii et al. (1937). The latter workers found that at low temperatures (20°C) reduction took place in two stages. At higher temperature (40°C–60°C) used by Marshall and Peters this effect is not evident. The sensitivity of the reaction rate to temperature was found by them to be virtually constant over a wide range of compounds. This was found to be true also of the slight but real sensitivity of the rate to reducing agent concentration. These two observations suggest that diffusion across the particle/water interface by the reagents is likely to be of some importance as already discussed.

Interest in the vatting equilibrium has been more widespread. This is because of the need for dye makers to make reasonably compatible vat dyes and also the need of dyers to use reducing agents efficiently because of their high cost. As with other reversible reducing systems vats exhibit an oxidation/reduction potential when a bright platinum electrode is put into the solution. This is because electrons from

$$> =0 + e \rightleftharpoons \!\!\!\!> \!\!\!-\, 0^-$$

are taken up. The electrode potential relates to the relative stabilities of the oxidized and reduced forms and this may be represented schematically.

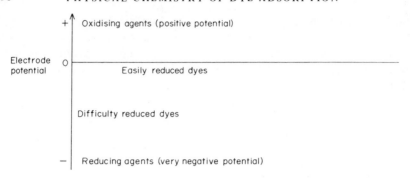

The reduction potential is related to the state of the equilibrium by,

$$\psi = \psi_0 + \frac{RT}{nF} \ln \frac{[D_0]}{[D_r]} - \frac{RT}{F} \ln [H^+] \qquad (6.38)$$

$[D_0]$ and $[D_r]$ are the concentrations of the quinone and leuco forms of the dye, n is the number of electrons involved. The other terms have their usual significance. The potential ψ_0 is a standard reduction potential equal to the observed potential when $[D_0] = [D_r]$ and $[H^+]$ is molar, i.e. pH 0. ψ_0 thus represents a standard state property suitable for comparing different compounds. As is to be expected from equation (6.38) the titration of a solution of a reduced soluble quinone with an oxidizing agent takes a sigmoidal form. If both quinone groups have the same reduction potential then the curve will take the form shown in Fig. 6.4. (solid line) when the two groups differ then semiquinone formation takes place (broken line).

Considering the regular reduction/oxidation titration curve, the central position can be seen to be linear and the mid-point is the 50% oxidation point, i.e. where $[D_0] = [D_r]$. The potential at that point, ψ mid is related to the standard reduction point by

$$\psi_0 = \psi \, \text{mid} - \frac{2 \cdot 303 \, RT}{F} \cdot \text{pH} \qquad (6.39)$$

deduced from equation (6.39).

The broken (semiquinone) curve is not unusual and is observed when in the two-stage reduction, the first is the rate determining step. Reduction is normally two stage but the rate determining effect is not always apparent.

The behaviour described relates only to soluble quinone and leuco compounds and consequently applies to vat dyes only under very dilute conditions in water which are not susceptible to experimental examination. When the quinone is precipitated due to its low solubility as in the case of a vat dye, the solution is rapidly saturated with the quinone form and $[D_0]$ becomes a constant. Equation (6.38) is not obeyed as a consequence and definition of the part of the titration curve prior to precipitation is difficult due to supersaturation effects as the vat dye precipitate nucleates.

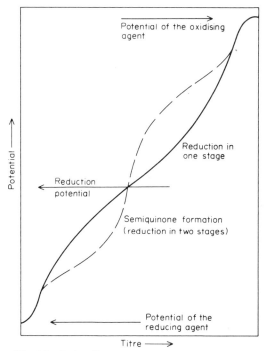

Fig. 6.4.   Redox filtration of quinonoid compounds.

Two theoretically based approaches have been taken towards a resolution of this difficulty and several empirical ones designed to assist dyers in practice. The theoretically based approaches are either to use polar solvents in which the quinone form of the dye has adequate solubility or to deduce a reduction potential from the small part of the titration curve which obeys equation (6.38).

Gupta (1952) was able to determine standard reduction potentials for several simpler anthraquinones in anhydrous pypridine solution. The pH was defined by supplying HCl gas to the solvent. Oxidation was carried out using ferric chloride in anhydrous pyridine and the potential was measured with reference to a silver-plated platinum electrode in anhydrous pyridine containing silver nitrate. The titration curves took the expected form and reduction potentials between $-500\,mV$ and $-70\,mV$ were observed. Gupta's data were later extended by Marshall and Peters (1952)(b). These investigations showed that even within the group of commercial vat dyes included in the compounds studied by Gupta and by Marshall and Peters, reduction potential varied widely. The actual value of the potential reflects the relative stabilities of the quinone and leuco forms. Using resonance theory this can be calculated (Carter, 1949). Gupta found that although the

calculated reduction potentials of the eight carbocyclic quinones studied
were numerically different from the experimental values there was good
correlation between them as shown in Fig. 6.5.

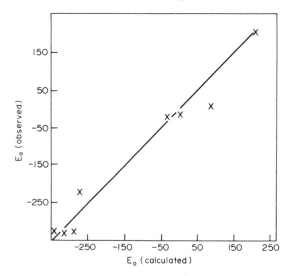

Fig. 6.5. A comparison of the experimental data for E_0 (Gupta, 1952) with predicted values
(Carter 1949).

Carter's formula which was used by Gupta can be applied only to simpler
molecules so that a more extensive evaluation of the possibility of calculating reduction potentials was not possible.

Due to the possible intrusion of solvent effects and also the need of dyers
for a method of following vat reduction under practical (i.e. aqueous) conditions efforts have persisted towards the achievement of useful results even if
they were not always theoretically justifiable. Geake and Lemon (1938)
investigated the use of 50% aqueous pyridine as a solvent but found that
truly reversible reduction was to be found with only simpler anthraquinone
derivatives due to precipitation. In a later investigation Appleton and Geake
(1941) measured the potential of the half oxidized system in the presence of
precipitated quinone. Among other solvents simple aqueous systems were
examined. If the solubility of the quinone is known the value of the standard
reduction potential is readily calculated. However if the solubility of the
quinone is treated as a small constant and if the leuco dye concentration at
the reference point is fixed, then the potential measured by Appleton and
Geake may be used for the purposes of comparison. In comparing two dyes
the difference between the two mid-point potentials will be

$$\Delta\psi_{mid} = \Delta\psi_0 + \frac{RT}{nF} \ln \frac{[Q_1]}{[Q_2]} \qquad (6.40)$$

where $[Q_1]$ and $[Q_2]$ are the solubilities of the two quinones. The error due to $[Q_1] \neq [Q_2]$ is, in practice, not large and meaningful comparisons may be made.

More recent investigators have been concerned with purely practical methods of comparing vat dyes and determining whether a vatting system is capable of maintaining a leuco dye solution. Schaeffer (1949) measured the potential associated with the reduction of the dye in the presence of oxygen, i.e. in a vessel open to the air. He recorded his results in terms of rH values. By analogy with pH, a low rH meant a strongly reducing solution was necessary to reduce the vat dye. Other investigators used modifications of the approach of Appleton and Geake in which they followed the titration of the leuco dye solution spectrophotometrically and were thus able to detect an end point when quinonoid dye began to be formed. In effect this approach is a refinement of Schaeffer's. The potential of a platinum electrode can be used in a purely practical sense as an indication of the state of the vat, warning the dyer that more reducing agent is needed. In combination with a pH meter this can be a very valuable production control system.

Although all the various methods of measuring some characteristic potential of a vat system give different numerical results, they do in general put the dyes in the same order of "vattability" and this corresponds with practical experience despite the fact that the conclusions of practical dyers are based as much on reduction kinetics as equilibria. Thus the reduction potential determined in some standardized fashion represents a useful expression of anticipated behaviour and is of value to the research chemist. Where there are discrepancies between different methods these are ascribable to solvent effects which can be quite large. An interesting comparison is given in Table 6.4.

TABLE 6.4.   A comparison of reduction potentials obtained by different methods.

Quinone	Mid point potentials (Appleton and Geake, 1941)		Standard reduction Potentials (Gupta, 1952; Marshall and Peters, 1952) in anhydrous pyridine
	in water	in 50% aqueous pyridine	
Flavanthrone	$-233\,mV$	$-391\,mV$	$-230\,mV$
3:4:8:9 dibenzpyrene 5:10 quinone	$-487\,mV$	$-340\,mV$	$-242\,mV$
Dimethoxy dibenzanthrone	$-575\,mV$	$-370\,mV$	$-264\,mV$
Dibenzanthrone	$-640\,mV$	$-328\,mV$	$-222\,mV$
Pyranthrone	$-660\,mV$	$-449\,mV$	$-300\,mV$

The solvent effect of pyridine is quite clear in two cases. In the case of dibenzanthrone and 3:4:8:9 dibenzpyrene 5:10 quinone it causes a marked fall in the potential so as to effect the ranking in the first instance.

The topic of vat dye reduction has not been studied since the 1950's despite the continued importance of the range. It would seem to have become accepted that the complications are too great for any meaningful advance on the position established at that time. The physical chemistry of the reducing agents has continued to arouse attention partly due to its intrinsic interest and partly due to the fact that the consumption of reducing agents represents a major cost factor in vat dye application. Two reducing agents are used sodium dithionite and sodium formaldehyde sulphoxylate. Additionally in the USA sodium borohydride is used and recently in an attempted economy, in India mixtures of sodium dithionite and sodium sulphide have been introduced. These compounds are not reversibly oxidized and the potential they develop at a platinum electrode is not strictly a reduction potential. Nevertheless the potential reflects their reducing power. With sodium dithionite the potential developed shows little change with temperature whereas with sodium formaldehyde sulphoxylate the potential at ambient temperatures is low, rising to the same value as that given by sodium dithionite by the time the temperature rises to 100°C. Thus the latter reducing agent is used in printing giving stable print pastes under room temperature storage and rapid reduction in the steamer.

The chemistry of sodium dithionite is complex and has been extensively studied. Polarography provides a particularly attractive technique for the selective study of the different reacting species. Three polarographic waves are shown at different temperatures (Broadbent and Peter, 1966; Milicevic and Eigenmann, 1963; Peter, 1956) as shown in Fig. 6.6. The first wave (A) occurs at a half wave potential of -0.5 V and is not markedly dependent on temperature. The second and third waves, (B) and (C), are very dependent on

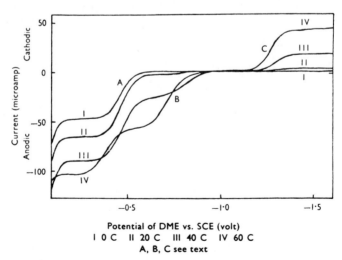

Fig. 6.6. Polarographic titration of sodium dithionite (reproduced with permission from Broadbent and Peters (1956)).

temperature. The wave (A) is also largely independent of pH and dithionite concentration and has been shown to be diffusion controlled, i.e. by the rate at which ions can diffuse to the electrode. The waves (B) and (C) are kinetic i.e. controlled by the progress of chemical reactions yielding the electroactive species.

The first wave (A) appears to be due to direct oxidation of dithionite ions to sulphite (Kolthoff and Tamberg, 1958). The second and third waves, (B) and (C), are thought to be due to the oxidation and reduction of the thionite radical ion produced by homolytic fission of the dithionite ion in alkali.

Thionite radicals have been shown to exist in solid sodium dithionite by ESR (Milicevic and Eigenmann, 1963). The reaction of oxygen with sodium

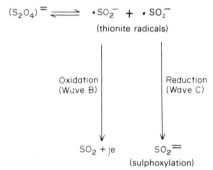

$$(S_2O_4)^{=} \rightleftharpoons \cdot SO_2^- + \cdot SO_2^-$$
(thionite radicals)

Oxidation (Wave B)   Reduction (Wave C)

$SO_2 + e$ $SO_2^{=}$
(sulphoxylation)

dithionite appears to involve thionite radical ions (Rinker *et al.*, 1960) as does the reduction of azo compounds by sodium dithionite (Wasmuth *et al.*, 1965). The thionite radical ($\cdot S_0^-{}_2$) is a very powerful reducing agent and is probably the major reducing species in vat dye reduction.

(3) *The adsorption of leuco vat dyes*

At one time it was considered that hydrogen bonding was of major importance in determining dye adsorption by cellulose. Consequently the apparently high affinity of leuco vat dyes without hydrogen bonding substituents constituted something of a mystery which triggered off an extensive research investigation by numerous investigators. While it is easy now to explain the adsorption behaviour of leuco vat dyes in terms of π–H bonding, dispersion forces etc., it must be accepted that to some extent this is due to the earlier investigations.

Due to the high dyeing pH extensive ionization of hydroxyl groups is to be expected in cellulose in vat dyeing and consequently low affinity values are to be anticipated. This was found to be the case, the values being at a level less than half of that observed with direct dyes on cellulose at pH 7 or thereabouts. The high substantivity was found to be due to the normally high ionic strength. A number of interesting structural features were also observed.

Lukin (1934) observed that the position of the negative charge centres in the leuco dye were important as can be seen from the following examples.

Leuco perylene   Leuco pyranthrone   Leuco isoamphipyranthrone

Low affinity   High affinity   Low affinity

Also in an unpublished study by Peter and Sumner the behaviour of substituted anthraquinones was examined revealing several sensitive structure/affinity relationships. Co-planarity was also established to be important in determining affinity.

Arising from the interest in the soaping stage of the vat dyeing process there was an extensive study of the steric disposition of absorbed vat dye molecule relative to the cellulose chain, (Sumner *et al.*, 1953). It was found that in a large number of cases the dyes were adsorbed with their main electronic absorption axis parallel to the fibre axis. This observation taken with the marked steric dependence of the affinity is a strong indication of the importance of dispersion forces in binding dependent as these are on a close proximation of the interacting species.

(4)   *The oxidation of leuco vat dyes*

This is a stage in the dyeing process which has been relatively little studied. Traditionally oxidation was carried out by hanging the dyed material in the air. Oddly enough this method is more likely to give rise to problems than the modern use of oxidizing agents such as potassium dichromate or sodium perborate in short treatment times.

Slow oxidation of leuco compounds creates several possibilities for side reactions.   The oxidation of sodium dithionite liberates sulphurous acid which leads to a lowering of the pH and the formation of the "acid leuco" compound, the hydroquinone. This can lead to further reaction to yield oxanthrones and anthrones. These may be of different colour from the desired quinone leading to off shade or weak results. This situation is shown schematically (p. 213) for 1:4 dibenzoylamino anthraquinone (C.I. Vat Red 42). An unsymmetrical leuco compound, e.g. 1 benzoylamino anthraquinone, can give two isomeric anthrones.

In general the equilibrium between the hydroquinone and oxanthrone forms is far in favour of the former. In the case where oxidizing agents are applied directly to the alkaline leuco dyed material the oxidation to the quinone is sufficiently rapid to make the formation of undesirable by-products negligible. Thus the demand for rapid oxidation procedure has obviated the problem which can arise in traditional manufacture.

On the other hand the use of oxidizing agents raises the possibility of over oxidation in particular cases leading to partial decomposition of the dye and consequent shade changes. Such factors are readily avoided by experience and appropriate dye selection procedures.

(5)   *The soaping of vat dyed materials*

After oxidation any vat dyed fabric onto which exhaustion has been less than 100% will contain surface quinonoid pigment from the dye in entrained solution. This will not rinse out easily and a light scouring treatment is necessary to achieve full fastness to rubbing. However in the majority of cases this after-soaping treatment performs an additional and at least as important task. Frequently soaping produces a change of shade in the dyed and oxidized fabric which is permanent. The dyeing thus is rendered faster to washing by the consumer. As a secondary effect fastness to chemical agencies, i.e. oxidizing agents, light etc., is also improved in some cases. It is clear that the dye pigment in its initial oxidized state is in a less stable condition than after soaping. Consideration of the causes must take into account molecular scale effects since the adsorption of the dye by the fibre occurs at that level and it is probable that the dye immediately following upon oxidation is in a molecularly dispersed condition on the fibre surface.

Early investigations of the events occurring during soaping led to the conclusion that crystal nucleation and growth occurred. The development of microscopically observable crystals within the fibre was noted in several cases whether soaping or steaming was used in the treatment (Bean and Rowe, 1929; Haller 1950, 1951). Bean and Rowe showed that very prolonged steaming could produce a decrease in rubbing fastness due to the growth of crystals on the actual fibre surface in some cases. Valkó (1941) carried out a much more extensive investigation of soaping using heavy shades of vat dyes applied to cellophane sheet. He examined the film by X-ray diffraction before and after soaping and could detect after long treatments the X-ray diffraction pattern of dye crystals superimposed on that of cellulose. The dyes examined fell behaviourally into three categories.

(a)   non-crystalline before and after soaping
(b)   non-crystalline before but crystalline after soaping
(c)   crystalline before and after soaping

The last observation, i.e. crystallinity before soaping, is interesting since it implies that nucleation either occurred during dyeing due, presumably, to stacking of the leuco dye or occurred very rapidly during the oxidation stage itself. The special behaviour of dyes in this group has not been particularly studied as far as is known.

A much more extensive study by Sumner et al. (1953) employed the dichroic nature of the vat dyes to follow changes in orientation of dye molecules on soaping. The absorption of light by the dye molecule is due to electronic oscillations induced in particular direction within the molecule. The energy of the light absorption depends upon the electronic characteristics of the oscillator in the direction in question. Involved in this process are the same delocalized (Π) electrons in the molecule which contribute to the development of dispersion force interactions. If the adsorbing substrate is a linear polymer which exhibits a general orientation as in the case in fibrous polymers, then the development of dispersion force interactions must lead to a general orientation of the adsorbed molecules with the primary oscillator parallel to the general orientation of the adsorbate molecules. Any movement of molecules in such a system to form say, crystallites in fibre voids, would cause a change in orientation which would be detectable as a change in the dichroic characteristics.

The dichroic effect of dye molecules adsorbed on cellulose fibres is fairly easily demonstrated. Adsorption of polarized light in the plane perpendicular to the fibre axis is very much less than it is when the plane of polarization is parallel to the fibre axis. It should be stressed that the orientation of dye molecules in a cotton fibre is a statistical matter. The cellulose chains are only orientated in general in the direction of the fibre axis so that the concept of orientation to the fibre axis includes all orientations at an angle of less than 45° to the fibre axis.

Sumner, Vickerstaff and Waters produced data for some thirty-two vat dyes which are shown in Table 6.5.

TABLE 6.5.   % orientation of vat dyes on viscose film before and after soaping at 100°C (Sumner *et al.*, 1953).

Vat dye	Time of Soaping (hrs)				
	0	0·5	5	8	
Caledon Yellow GN	0	16⊥			
Caledon Yellow 5 GN	7	13⊥	16⊥		
Caledon Brilliant Orange GR	0	7⊥			
Caledon Gold Orange G	0	22⊥	22⊥		
Caledon Red BN	0	13⊥			⊥ = "perpendicular"
Caledon Red 2G	0	22⊥			orientation
Caledon Red X5B	0	7⊥			
Caledon Brilliant Violet R	0	7⊥			∥ = "parallel"
Caledon Violet XBN	7	13⊥			orientation
Caledon Green 7G	0	7⊥			
Caledon Yellow 5G	7	7⊥	28⊥		
Caledon Brilliant Red 3B	7	7⊥			
Caledon Green RC	4∥	7⊥	7⊥		
Caledon Green 2B	7∥	7⊥			
Caledon Blue RC	4∥	13⊥			
Caledon Brilliant Blue 3G	4∥	7⊥			
Caledon Brilliant Blue RN	4∥	13⊥	22⊥		
Caledon Yellow 2C	7∥	4∥	7⊥		
Caledon Brilliant Orange 4RN	4∥	0			
Caledon Dark Brown 2G	4∥	0	7⊥		
Caledon Brilliant Purple 4R	4∥	0			
Caledon Jade Green 2G	7∥	4∥	13⊥		
Caledon Dark Blue 2R	13∥	0	22⊥		
Caledon Yellow 4G	7∥	7∥	0	7⊥	
Caledon Orange 2RT	0	0	16⊥		
Caledon Gold Orange 3G	7∥	7∥	0	7⊥	
Caledon Brown 3G	4∥	4∥	2∥	2⊥	
Caledon Yellow 3G	0∥	19∥			
Caledon Red 5G	4∥	19∥			
Caledon Pink RL	0∥	7∥			
Caledon Brilliant Violet 3R	7⊥	0	4∥		

The values for the degree of orientation reflect the fact that the substrate is far from completely oriented as already stated. In order to calculate a degree of orientation Sumner *et al.* accepted the convention that if the optical density of the dyed film measured with plane polarized light with the electric vector parallel to and perpendicular to the direction of extrusion of the film giving $D\|$ and $D\perp$ respectively, then when $D\| > D\perp$ the dye is oriented mainly in the parallel direction. The fractional orientation is given by

$$\frac{D\| - D\perp}{D\| + 2D\perp} \tag{6.41}$$

On the other hand when $D\| < D\perp$ the fractional perpendicular orientation is given by

$$\frac{2(D\perp - D\|)}{D\| + 2D\perp} \tag{6.42}$$

The factors for 2 applied to D and $(D\perp - D\|)$ take into account the assumption of random radial distribution of dye around the axis of extrusion.

The results can be seen to reflect Valkó's data. The dyes fall into five groups:

(a) dyes changing from random to perpendicular orientation within 30 minutes and increasing in that orientation with time

(b) dyes changing from parallel to perpendicular orientation within 30 minutes

(c) dyes changing very slowly from parallel to perpendicular orientation

(d) dyes changing extremely slowly to perpendicular orientation

(e) dyes remaining in or turning to parallel orientation.

The classification is quite arbitrary but reflects the spectrum of behavioural characteristics of the vat dyes examined in this way. It must be noted that the orientation changes are slow in all cases and apart from group (e) involve movement to perpendicular orientation. Sumner *et al.* formed the conclusion that the vat dyes were forming microcrystalline needles in voids in the substrate which ran parallel to the direction of extrusion and that the dye molecules were arranged perpendicular to the crystal axis. At that time only one vat dye crystal had been examined and this had such a configuration. The tendency to parallel configuration immediately following oxidation is not unexpected in view of the adsorption mechanism. The overall conclusion drawn was that "the unsoaped dyeing is merely an arbitrary stage in the process of crystallization which begins immediately after oxidation and is completed by soaping".

Orientation changes as described by Sumner *et al.* have been recently confirmed by Gulbrajani (1971) using attenuated total reflectance polarized infra red spectroscopy in which the movement of the carbonyl group in the oxidized dye was followed. This is an unambiguous method and fully confirms that the dye molecules may rotate in soaping. In view of the weight of evidence in favour of the crystallization theory it may seem surprising that it has been rejected by some investigators. Their views have been presented by Wegmann (1960) and are based on a number of observations which do not fit in easily with a simple crystallization theory. The main points are as follows:

(a) in some cases there is no correlation between the dichroism of the soaped dyeings and the various known crystalline forms of the dye

(b) shade changes similar to those produced by soaping may be produced in solutions of vat dyes in organic solvents when methanolic potassium hydroxide is added

(c) colour changes produced slowly in soaping at 100°C may be produced almost instantaneously, in some cases, by treatment in cold water/polar solvent mixtures

(d) in some cases soaping will produce very different shades on different forms of cellulose

(e) in some cases the shade change produced depends on the history of the dyeing after oxidation.

Of these points the most interesting are (c), (d) and (e) which provide evidence of new factors which need to be taken into account in any crystallization theory. Wegmann argues that if molecules must move through a 90° angle to adopt a crystalline configuration it is unlikely to occur at 25°C in water/polar solvent mixtures in a medium as constrained as a cellulosic fibre or film. This is a reasonable point to make although the evidence that molecules *do* make such a shift in aqueous soaping is very strong. The only way to resolve these observations is to separate the question of the shade change from that of crystal growth as will be seen. The effect of different substrates and soaping history are exemplified below.

Dye	Fibre	Oxidized unsoaped shade	Soaped shade
	Cotton	yellowish red	red
	Rayon	yellowish red	bluish red
	Cotton	brownish red	yellowish red
	Rayon	brownish red	violet

The effect of soaping history is shown very clearly in the following example.

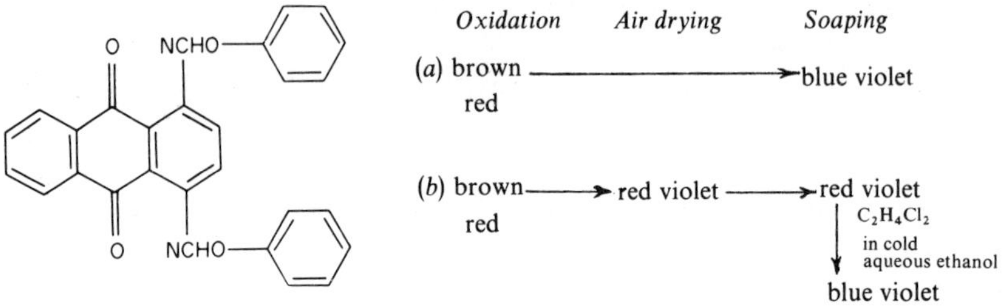

on cotton

Wegmann argues that in the adsorption of the leuco form of the dye, the development of dispersion force interactions results in the adoption of a particular molecular configuration involving dye and substrate. On oxidation this configuration may or may not be disturbed but it represents a constraint which means that the molecular configuration of the adsorbed quinone form is metastable. The shade development according to Wegmann involves the adoption by the molecule of a more stable spacial arrangement and since this will involve shifts in the electric oscillators results in a shade change. The main oscillator can shift through 90° giving a dichroic effect without movement of the molecule as a whole. Wegmann describes these forms of an adsorbed vat dye as tactile isomers and a typical hypothetical transition is shown in Fig. 6.7 below.

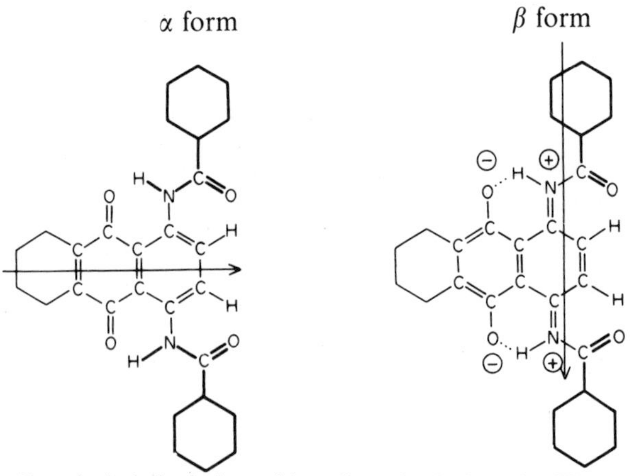

Fig. 6.7.   Hypothetical physical transition of an adsorbed vat dye (Wegmann, 1960).

The heavily marked parts of the two configurations are out of the plane of the paper. The α form is one of the possible configurations of the unsoaped form while the β form is one of the possible soaped configurations. The arrow indicates the direction of the main electronic oscillator. Recent work by Jones and Arnold (1971) on thermal transitions in vat and other dye crystals has shown that a number of low energy transitions with associated colour changes occur with many dyes in the solid state and that no change in crystal habit, such as might be detected by X-ray diffraction, occurs. They have found that dyes in the solid state almost universally exhibit many transitions to forms of different stability. Thus changes such as are discussed by Wegmann are known to be possible and could explain colour changes which occur on soaping. However the number of crystalline forms which are possible with dyes observed by Jones suggest that the stable crystal habit within a cellulose fibre, for example, need not be that observed outside the fibre and could indeed vary with the structure of the cellulosic fibre itself, e.g. as between cotton and viscose rayon. In addition the growth of microcrystals within a fibre could cause colour changes by light scattering and multiple reflectance effects to judge from the work of Hannam and Patterson (1963).

It would seem that the question of configurational transitions and crystal growth are inseparable in the consideration of soaping vat dyes. Wegmann's contention that crystal growth is not a factor is likely to be an over simplification as would be any hypothesis based on the observation of crystal growth alone. Solid state transitions in pigment crystal has been for many years recognized as an important topic. It is only recently that the importance of such changes in dye solids has begun to be recognized. The development of vat dyes by soaping after oxidation is one aspect only of this wider issue.

IV   REFERENCES

Allingham, M. M., Giles, C. H. and Neustadter, E. L. (1954) Discussion of the Faraday Society, No. 14.

Appleton, D., and Geake, A. (1941) *Trans. Far. Soc.*, **37**, 45.

Aspland, R., and Bird, C. L. (1961) *J. Soc. Dyers and Colourists*, **77**, 9.

Bader, M. (1924) *Chim. et. Ind.*, **455**, 449.

Bader, M. (1938) *Amer. Dyes Rep.*, **27**, 455.

Barrer, R. M. and Wasilewski, S. (1961) *Trans. Far. Soc.*, **57**, 1140.

Bean, P. and Rowe, F. M. (1929) *J. Soc. Dyers and Colourists*, **45**, 67.

Belenkii, L. I. Sokolov, I. I. and Kazanskaya, M. E. (1937) *Khlopchatobumazhnaya Prom.*, **7**, 32.

Broadbent, D. A. and Peter, F. (1966) *J. Soc. Dyers and Colourists*, **82**, 264.

Carrol, B. and Cheung, H. C. (1962) *J. Phys. Chem.*, **66**, 2585.

Carter, P. G. (1949) *Trans. Far. Soc.*, **45**, 597.

Crank, J. (1947) *J. Soc. Dyers and Colourists*, **63**, 293.

Daruwalla, E. H. and D'Silva, A. P. (1963) *Textile. Res. J.*, **33**, 40.

Devanathan, M. A. V. (1962) *Proc. Roy. Soc.*, **267**, 256.

Dundon, M. L. and Mack, E. (1923) *J. Amer. Chem. Soc.*, **45**, 2479.

Evans, H. C. (1958) *J. Colloid Sci.*, **13**, 537.

Farrar, J. and Neale, S. M. (1952) *J. Colloid Sci.*, **7**, 186.

Fowler, J., Michie, A. G. H. and Vickerstaff, T. (1951) *Melliand Textilber*, **32**, 296.

Geake, A. and Lemon, J. T. (1938) *Trans. Far. Soc.*, **34**, 1409.

Graham, R. P. and Fromm, H. J. (1947) *Canad. J. Res.*, **25**, 303.

Gulbrajani, M. L. (1971) Ph.D. Thesis University of Bombay.
Gupta, A. K. (1952) *J. Chem. Soc.*, 3473, 3579.
Haller, R. (1950) *Helv. Chim. Acta.*, **33**, 1165.
Haller, R. (1951) *Helv. Chim. Acta.*, **34**, 793.
Halfpenny, E. and Robinson, P. L. (1952) *J. Chem. Soc.*, 928, 940.
Hannam, A. R. and Patterson, D. (1963) *J. Soc. Dyers and Colourists*, **79**, 192.
Hanson, J. and Neale, S. M. (1934) *Trans. Far. Soc.*, **30**, 386.
Hanson, J., Neale, S. M. and Stringfellow, W. A. (1935) *Trans. Far. Soc.*, **31**, 1718.
Iyer, S. R., Srinivasan, G., Baddi, N. T. and Ravikrishnan, M. R. (1964) *Text. Res. J.*, **34**, 807.
Iyer, S. R. and Baddi, N. T. (1968) *Contribution to the Chemistry of Synthetic Dyes and Mechanism of Dyeing*, Symposium proceedings, University of Bombay.
Iyer, S. R., Srinivasan, G. and Baddi, N. T. (1968) *Text. Res. J.*, **38**, 693.
Iyer, S. R. and Singh, G. S. (1969) *Physico chemical aspects of the interaction of dyes in solution and in fibre systems*, Symposium proceedings, University of Bombay.
Iyer, S. R. and Jayaram, R. (1970) *J. Soc. Dyers and Colourists*, **86**, 398.
Jeffries, R. (1960) *J. Text. Inst.*, **51**, T339.
Johnson, A. and Ainsworth, S. (1955) *J. Soc. Dyers and Colourists*, **71**, 592.
Johnson, A. and Ainsworth, S. (1957) *J. Soc. Dyers and Colourists*, **73**, 41.
Johnson, A. and Rahman, H. L. (1958) *J. Soc. Dyers and Colourists*, **74**, 291.
Johnson, A. and Lockett, A. P. (1960) *J. Soc. Dyers and Colourists*, **76**, 412.
Jones, F. and Arnold, P. (1971) Unpublished research, University of Leeds.
Knapp, L. F. (1922) *Trans. Far. Soc.*, **17**, 457.
Kolthoff, I. M. and Tamberg, N. (1958) *J. Polarographic Soc.*, **1**, 54.
Lukin, A. M. (1934) *Anilinokras. Prom.*, **4**, 536.
Marshall, W. J. and Peters, R. H. (1947) *J. Soc. Dyers and Colourists*, **63**, 446.
Marshall, W. J. and Peters, R. H. (1952) (a) *Bull. Inst. Text. France*, **30**, 415. (b) *J. Soc. Dyers and Colourists*, **68**, 289.
Milicevic, B. and Eigenmann, G. (1963) *Helv. Chim. Acta.*, **46**, 192.
Neale, S. M. and Stringfellow, W. A. (1940) *J. Soc. Dyers and Colourists*, **56**, 17.
Nishida, K. (1951) *J. Soc. Text. Ind. Japan*, **7**, (539)
Nishida, K., Akimoto, T. and Uedira, H. (1966) *Kolloid Zeit.*, **233**, 896.
Peter, F. (1956) *Acta. Chim. Acad. Sci. Hungary*, **9**, 421.
Rattee, I. D. and Seltzer, I. (1950) Unpublished research.
Robinson, C. (1954) Discussions of the Faraday Society, No. 14.
Rinker, R. G., Gordon, T. P., Mason, D. M., Sakaida, R. R. and Corcoran, W. H. (1960) *J. Phys. Chem.*, **64**, 573.
Schaeffer, A. (1949) *Melliand Textilber*, **30**, 111.
Sumner, H. H., Vickerstaff, T. and Waters, E. (1953) *J. Soc. Dyers and Colourists*, **69**, 181.
Tamamushi, B. and Tamamatu, K. (1959) *Trans. Far. Soc.*, **55**, 1007.
Urquhart, A. R. and Williams, A. M. (1924) *J. Text. Inst.*, **15**, T138; T433; T559.
Valko, E. I. (1941) *J. Amer. Chem. Soc.*, **63**, 1433.
Verwey, E. J. W. and Overbeek, J. T. D. (1948) *Theory of the stability of hydrophobic colloids*, Elsevier.
Vickerstaff, T. (1954) *The Physical Chemistry of Dyeing*, 2nd ed., Oliver and Boyd, London.
Wasmuth, C. R., Donnell, R. L., Harding, C. E. and Shankle, G. E. (1965) *J. Soc. Dyers and Colourists*, **81**, 403.
Wegmann, J. (1960) *J. Soc. Dyers and Colourists*, **76**, 282.
Willis, H. F., Warwicker, J. O., Urquhart, A. R. and Standing, H. A. (1945) *Trans. Far. Soc.*, **41**, 506.

Chapter 7

The adsorption of non-ionic disperse dyes

I GENERAL

Attention has thus far been confined to dyes which possess relatively high solubility in water and are applied to fibres etc. in the ionized state. However man-made polymeric substrates, e.g. polyester, polypropylene, cellulose di- and tri- acetate, contain a very low concentration of ionic groups and, although they adsorb significant amounts of water, compared with naturally occurring polymers they are hydrophobic. This has the effect of greatly increasing electrical repulsion effects when ions are adsorbed. In addition the low water adsorption of these polymers reflects low swelling and consequently they are difficulty permeable by all but small dye molecules. In order to dye such fibres (represented in the 1920's by cellulose diacetate alone) weakly polar dyes of relatively low molecular weight were developed. Their lack of ionic solubilization led to sparing aqueous solubility and when they are applied in aqueous systems they are present initially as dispersions and it is from this that the generic term disperse dyes has arisen. Their dyeing behaviour has given rise to much theoretical discussion regarding the way in which they become adsorbed, i.e. transfer from the crystalline to the adsorbed state, and also the molecular mechanism of the dye-fibre interaction itself. The physical chemical understanding of disperse dye systems still leaves a very great deal to be desired and this situation seems likely to continue as new aspects of the problem are still being revealed.

II THE TRANSFER OF DYE FROM THE CRYSTALLINE TO THE ADSORBED STATE

In the application of a disperse dye the substrate, e.g. cellulose diacetate material, is immersed in a dispersion of the dye in water and dyeing ensues. Two views of the mechanism of transfer are possible. Either solid material becomes adsorbed onto the substrate surface followed by a solid state diffusion process or dye is transferred by normal processes from the very dilute solution surrounding the dispersion. Both views were expressed in the early days of disperse dyes. The second (the aqueous solution theory) was put forward by Clavel and Stanisz (1923, 1924) without experimental evidence while the first view was expounded by Kartaschoff (1925) and was apparently supported by material evidence. Kartaschoff observed particles of crystalline dye adhering to the surface of cellulose acetate fibres and found that these disappeared apparently by adsorption at 60°C. In addition dry cellulose acetate in contact with dry disperse dye for several days at 60°C became dyed. This evidence combined with a general belief that disperse dyes had no solubility led to the general adoption of Kartaschoff's view. This was unfortunate because it was wrong.

Credence was given to the solid state transfer theory by the X-ray diffraction evidence of the 1920's which showed cellulose acetate to have no fine structure. It was regarded as a "Solid colloid". In addition it was considered that due to the failure of basic dyes to penetrate cellulose acetate despite the presence in it of carboxyl groups, that cellulose acetate contained no pores large enough to permit the entry of dye molecules (Valkò, 1937). Thus dye molecules penetrating cellulose acetate must, it was considered, diffuse through the "polymer substance" itself rather than through voids in it by some process of solid solution.

Vickerstaff and Waters (1942) failed to detect any attraction of dye particles to the fibre in their experiments. Lauer (1932) also failed to repeat Kartaschoff's experiment and even showed that dye particles normally were negatively charged so that they could not be taken up by the fibre. Later Millson and Turl (1951) also failed to substantiate Kartaschoff's results. Kartaschoff noted a positive charge on the particles used in his dyeing experiments so there is no doubt that he *did* observe solid dye adhesion. What other experimenters found in effect was that this was irrelevant to the question of dye transfer. This was confirmed by Vickerstaff and Waters when they successfully dyed a cellophane sheet in a dyebath where the solid particles were kept isolated in a dialysis bag. The matter was finally settled by the work of Bird, Manchester and Harris (1954), Bird (1954) and of Schuler and Remington (1954). Firstly it was shown by Bird that disperse dyes had appreciable solubility so that the ideas of Clavel expressed some thirty years before were quite feasible. Solubility data due to Bird, Manchester and Harris and to Schuler and Remington are shown in Table 7.1.

TABLE 7.1.   Solubility data for disperse dyes in water.

Dye	Temperature °C	Solubility (molar)
p-nitroaniline→aniline	80	3.93×10^{-5}
p-nitroanline→diphenyl	80	1.57×10^{-6}
p-nitroaniline→ N-ethyl-N-β-hydroxyethyaniline	80	2.23×10^{-5}
1:4 diamino anthraquinone	80	6.39×10^{-5}
1:4 diamino-2-methoxy anthraquinone	80	3.79×10^{-5}
1:4 dihydroxy anthraquinone	100	6.25×10^{-5}
Iamino 4 hydroxy anthraquinone	120	6.24×10^{-4}

Bird also showed that in many cases significant increases in solubility were produced by the addition of surfactants. Both groups of investigators demonstrated that the dyeing process was reversible so that equilibria were established by adsorption or desorption and conformed to a rectilinear partition isotherm ($[D_f]/[D_s] = K$). Bird, Manchester and Harris concluded that the dyes were binding (weakly) onto sites in the fibre by hydrogen bonding. Schuler and Remington on the other hand found that some dyes exhibited "additive" behaviour. This means that in a binary mixture of the dyes there is no evidence of competition so that each dye is adsorbed as if the other was absent. This led them to conclude that there were no adsorption sites but at the same time there was an "interaction between the dyes and the fibre, possible hydrogen bonding".

The transfer of dye from the crystalline to the adsorbed state through aqueous solution has been paralleled by the work of White (1960) who dyed cellulose acetate with disperse dyes from solutions in benzene and carbon-tetrachloride. Datye and co-workers (1971) have produced extensive data relating to dyeing from perchloroethylene. The work of Marjury (1956) established that dyeing could occur from the vapour phase without water being present thus explaining Kartaschoff's dry dyeing experiment. It is thus well established that the dyeing process with disperse dyes depends upon the achievement of molecular dispersion of the dye either in solution or in the vapour and that it is this state which is adsorbed. Due to the sparing solubility of the dyes and their low vapour pressure it is easy to achieve a practical upper limit to the external concentration and this inevitably limits the adsorption. This is illustrated in Fig. 7.1.

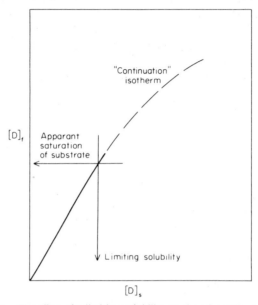

Fig. 7.1.   The effect of a limiting solubility on the adsorption isotherm.

This effect was first noted by Vickerstaff and Waters (1942) but has been recognized only recently as being of practical importance. In their investigation Vickerstaff and Waters were concerned with the possible effects of particle size on the dyeing behaviour of the dispersion. However they were unable to reach any very definite conclusions because their dispersion was not stable. When dyeings were prolonged the gradual growth of the particles caused a lowering of the solubility and this reduced the maximum dye uptake by the fibre. The effect is illustrated in Fig. 7.2.

In extremely fine dispersions the solubility of the particles may be higher than the normally observed solubility. The situation is analogous to that described by the Kelvin equation for the vapour pressure of a liquid in spherical drops,

$$\ln \frac{p_a}{p} = \frac{2\gamma v}{akT} \tag{7.1}$$

in which p_a is the vapour pressure over a surface of radius a and p is that over a flat surface; v is the molecular volume of the liquid, γ is the surface tension and kT is a Boltzmann energy term. The analogous equation for the solubility of particles in a dispersion is

$$\ln \frac{C_a}{c} = \frac{2\gamma_i v}{akT} \tag{7.2}$$

in which γ_i is the interfacial energy or tension, c and C_a are molar concentrations of dissolved material in contact with a solid/solvent interface which is planar and of radius a respectively. The term $2\gamma_i v/akT$ becomes very small as a is increased so that $C_a \rightarrow c$. However the effect of equation (7.2) is that small particles are more soluble than larger ones with

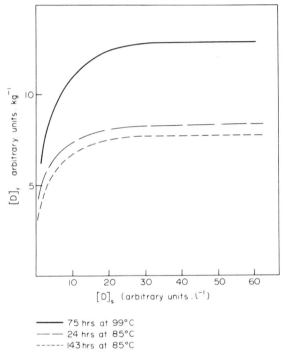

Fig. 7.2. The effect of prolonged dyeing on the adsorption isotherm of 1-isobutylamino-4-anilino anthraquinone (Vickerstaff and Waters, 1942).

the consequence that a solution may be supersaturated with respect to large particles when small particles are present. Thus crystallization occurs on the surface of large particles so that the mean size increases and the mean solubility falls. The effect of this in a disperse dyebath is to lower the limiting solubility and hence the optimum dye uptake by the substrate. This effect has become of great technical interest in the recent period since in an attempt to provide disperse dyes of high fastness for polyester fibres, disperse dyes of very low aqueous solubility have been developed. In order to achieve sufficiently rapid dyeings, temperatures of up to 140°C have been employed.

Under these conditions the rate of crystal growth has been in some cases sufficiently rapid for the solubility of the dye to fall more rapidly than it is adsorbed by the polyester. Thus large crystals of very low solubility have been formed resulting in very weak dyeings and poor rubbing fastness.

This technical effect has given impetus to the study of the solid state properties of dyes in general and disperse dyes in particular. Physical transitions under conditions of heating have been shown to be almost commonplace by Jones and Flores (1970, 1971) and Biederman (1971) has shown that with the dye

five crystal modifications can be isolated with different aqueous solubilities. The dyeing behaviour follows the same isotherm in each case but the optimum dye uptake by the fibre varies with the solubility. The "saturation values" are shown in Table 7.2.

TABLE 7.2. "Saturation" values obtained with the different crystal modifications of Dye 7.1 (Biedermann, 1971).

Physical modification	"Saturation" value (g/100g) dry fibre
α	0·77
β	1·00
γ	2·13
δ	3·07
ε	5·50

From a technical point of view it is perhaps unfortunate that the most soluble crystal modification is the least stable so that technical advantage is not easily taken of the situation. However, the kinetics of the transitions are of increasing interest since stability is required only for the duration of a dyeing operation.

III   THE NATURE OF THE DYE ADSORPTION

Consequent upon the ideas of Kartaschoff (1925), Valkó (1937) and others the view came to be adopted that disperse dyes were retained in the fibre by a process of "solid solution". Some of the reasons for this concept being adopted have been discussed. It has a certain plausibility encouraged by the observation of rectilinear adsorption isotherms in the majority of cases studied and the analogy between these and the partition of a solute between two immiscible solvents. As against this, the existence of adsorption sites is firmly accepted for dye ions applied to polyamide, protein or cellulosic substrates and this has led to certain reluctance to adopt a postulate which implies that reduction of the polarity of a dye causes a totally different mechanism to operate. Vickerstaff (1954) has pointed out that in any case the distinction between solid solution and site adsorption models must be very largely unreal. Solid solution must imply solid solvation, i.e. an association between the dye molecule and the polymer acting as a solvent which is indistinguishable from adsorption at a site. The only possible distinction which might be made would involve the presumption that the dye can penetrate what Vickerstaff calls "micellar parts" of the polymer which are distinct from pores or voids. There is no evidence that disperse dyes penetrate regions of high order in polymers or that they cause any gross change in the polymer morphology except in very unusual circumstances. It is reasonable to conclude as a consequence that the distinction between the two models does not exist. Nevertheless the concept that disperse dyes are adsorbed by some unusual mechanism has persisted. In part this has been due to the belief that the nature of a site must be defined before it can be postulated and since this is not easy it is often suggested that sites do not exist. However there is a considerable body of evidence that they do.

Early investigations by Marsden and Urquhart (1942) of the adsorption of phenol by cellulose acetate showed that at low uptake levels insufficient to cause fibre swelling phenol was taken up in accordance with a curvilinear isotherm. However when sufficient phenol was adsorbed to bring about significant disruption of cohesive forces in the polymer swelling occurred to a significant degree leading to further uptake. A comparison of the swelling and adsorption behaviour of phenol on cellulose acetate is shown in Fig. 7.3. taken from the data of Marsden and Urquhart (1942). Thus even so simple a substance as phenol demonstrably adsorbs onto sites at low concentrations and is not capable of penetrating freely into the whole of the polymer network. Due to their low solubility disperse dyes are absorbed to a much lower concentration level than is phenol and it appears unlikely that they will move more freely in the polymer since they have a much greater molecular size.

Investigations reported by Patterson (1954) strongly support specific binding of disperse dyes in polyethylene terephthalate and suggest location of the dye in amorphous regions of the polymer. The disperse dye C.I. Disperse Red I was applied to amorphous polyethylene terephthalate film which was then drawn to produce polymer chain orientation. At different draw ratios the dichroic ratio was determined with the result shown in Fig. 7.4.

The behaviour of the dye does not precisely follow the development of strain-induced crystallization as reflected in density and birefringence measurements as these are not sensitive to increases in order in the amorphous regions of the polymer below a fairly high degree. More recent studies by Blacker and Patterson (1969) have shown that disperse dyes are taken up by polyethylene terephthalate initially in regions of least order and that as the concentration of dye in the polymer increases regions in the polymer of increasing order become occupied. This effect strongly implies a close asso-

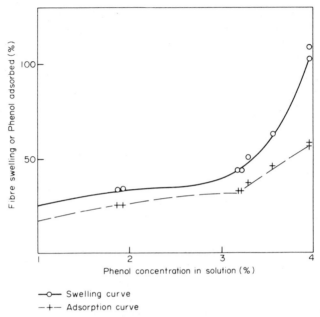

—o— Swelling curve
—+— Adsorption curve

Fig. 7.3.   A comparison of the swelling and adsorption behaviour of phenol on cellulose acetate (Marsden and Urquhart, 1942).

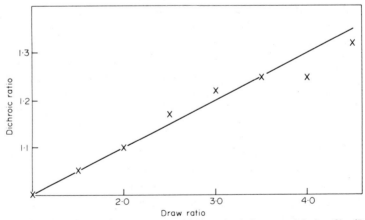

Fig. 7.4.   The orientation of C.I. Disperse Red I in polyethylene terephthalate film (Patterson, 1954).

ciation of dye molecules with polymer chains of the kind which cannot be described meaningfully as a solid solution.

Further evidence of a similar kind has been produced by Merian (1966) and Husy, Merian and Schetty (1969) as a result of the study of photomechanical effects in cellulose acetate. Three dyes 7.1, 7.2 and 7.3 below were all found to exhibit phototropisin on cellulose acetate film.

Dye 7.1

$$CH_3-O, CH_2-O \quad CH_3-CH-CH_2 \quad CO-NH-\langle\rangle-N=N-\langle\rangle-OH \quad (NHCOCH_3)$$

Dye 7.2

$$CH_3-O \quad CO-NH-\langle\rangle-N=N-\langle\rangle-OH \quad (NHCOCH_3)$$

Dye 7.3

$$CH_3-O, CH_2-O \quad CH_3-CH-CH_2 \quad CO-NH-\langle\rangle-N=N-\langle\rangle-OH \quad (NHCOC_4H_{19})$$

Dyes 7.1 and 7.2 were found to behave virtually identically but Dye 7.3 was much more resistant. This showed that the nature of the acyl.group in the diazo component did not make a great deal of difference while that on the coupling component could have a profound effect. The phototropism was ascribed to cis-trans isomerisin. The dye 7.4,

Dye 7.4

$$CH_3-O, CH_2-O \quad CH_3-CH-CH_2 \quad CO-NH-\langle\rangle-N=N-\langle\rangle(CH_3) \quad H\cdots O$$

was found to show no phototropic effect presumably due to internal hydrogen bonding. Phototropism had been observed earlier by Mechel and Stauffel (1941) using amino azo benzene derivatives adsorbed on cellulose acetate. They established in their case that cis-trans isomerism was involved and concluded that the dye molecules must be unusually free to move in the polymer so that no adsorption sites could be involved. This conclusion is consistent with their observation although it does not exclude site adsorption in fact. This was shown by Merian when Dye 7.1 was observed to exhibit photomechanical effects. A dyed strip of cellulose acetate was observed to shrink by 0·085% on exposure to light due to the smaller molecular length of the cis-form of the dye. Undyed cellulose acetate did not shrink on exposure and the effect with the dyed strip was reversible. Similar results were obtained on nylon and also with other dyes capable of cis-trans isomerism. A close conformational association of dye molecules and the substrate was indicated as a result of these experiments.

Studies by Daruwalla, Rao and Tilak (1960) have shown that steric factors are of much greater importance in connection with adsorption than in the case of solution. A series of dyes were prepared in which planar and non-planar molecular analogues were represented. Taking three pairs of such dyes, the "partition coefficients" describing the dyeing isotherms and the relative solubilities in water and amyl acetate were studied. The planar and non-planar analogues differed very little in their solubilities in amyl acetate, but the latter were in each case more soluble in water and showed much lower "saturation values" on cellulose acetate. This last effect was the most noticeable effect of inducing non-planarity in the dye molecule (see Fig. 7.5) and indicates that general binding forces are almost certainly involved.

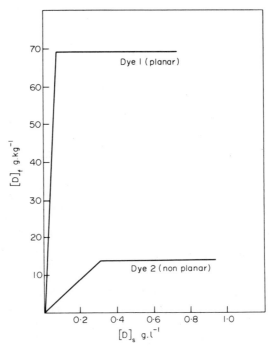

Fig. 7.5. Partition isotherms for planar and non-planar dye analogues applied to cellulose acetate.

The difference between dissolving the dyes in a solvent such as amyl acetate and "saturating" a cellulose acetate film with the dye is made more clear by a comparison of the solubilities of the dyes used by Daruwalla *et al.* in amyl acetate and the "saturation" values of the film. The effect of planarity is most clearly to be seen by comparing the solubilities of the two forms in amyl acetate, in water and in cellulose acetate. In Table 7.3 the ratios are shown for three pairs of dyes.

TABLE 7.3.   Solubility ratios (planar/non planar) for disperse dyes in different solvents (Daruwalla, Rao and Tilak, 1960).

Pairs of dyes (Daruwella's nomenclature)	Solvent		
	Amylacetate	Cellulose acetate	Water
I and II	1·58	6·83	0·27
III and IV	1·22	43·55	0·12
V and VI	1·32	43·55	0·58

It can be seen that the planarity factor has only a slight effect on the solubility ratios of the dyes in amyl acetate and in water. In general planarity aids solubility in amyl acetate and limits solubility in water. In cellulose acetate on the other hand the effect is very large indicating fairly strong interaction in the case of the planar dyes.

One of the main arguments presented against a site adsorption model has been the existence in so many cases of a rectilinear partition isotherm on several hydrophobic polymeric substrates. On the one hand this has encouraged analogies with partition between immiscible solvents and the argument that adsorbed dye in the polymer must have a freedom of movement akin to that in a solution. On the other hand a much more plausible and sophisticated argument is based on the fact that when adsorption is described by a partition isotherm there is no evident probability or site occupation factor so that site adsorption may be discounted. Any consideration of a site adsorption model cannot be seriously considered unless these factors are taken into account.

The analogy between the solvent-like behaviour of hydrophobic polymers and solvents such as ethyl acetate, amyl acetate etc. can be regarded as highly tenuous. Quite apart from the discrepancies in the analogy shown in Table 7.3, there remains the fact that disperse dyes give linear partition isotherms not only on polar substrates such as nylon but also on hydrophilic polar substrates such as wool (Bird and Firth, 1960).

The probability factor requires more serious consideration. The simplest argument put forward by proponents of the site adsorption model is that partition isotherm is the early part of a Langmuir isotherm. The Langmuir isotherm equation can be expressed in two forms,

$$[D_f] = K_{ads}[S][D_s]/1 + K_{ads}[D_s] \qquad (7.3)$$

and
$$[D_f]/[D_s] = K_{ads}([S] - [D_f]) \qquad (7.4)$$

in which $[S]$ is the true saturation value. Clearly under circumstances in which $K_{ads}[D_s] \ll 1$ or $[D_f] \ll [S]$ then $[D_f]/[D_s]$ will become a constant and the system will appear to follow a partition isotherm. The feasibility of the proposition that this occurs with disperse dyes can be calculated by seeing whether the necessary conditions apply in known cases.

Bird *et al.* (1954) report solubility values (and hence *maximum* values of $[D_s]$) with an order of magnitude of 10^{-3} g/l. Values of K_{ads} calculated on the same basis are of the order of 10^3 so that $K_{ads}[D_s]$ is always comparable with unity. In some cases $[D_s]$ is significantly smaller than 10^{-3} g/l but then K_{ads} tends to be higher in such circumstances.

Arguments based on the view that the maximum value of $[D_f]$ obtainable with disperse dyes is always much less than the saturation value have been put forward by Stubbs (1954) and H. J. White (1960). No experimental proof can be advanced but it is possible to assess the implications of this idea. In order for $[S] - [D_f]$ to experimentally approach $[S]$ then the value of $[D_f]/[S]$ must certainly be less than 0·1. Considering typical data (e.g. Seddon, 1962) maximum dye uptake values of 100 g/kg are quite possible so that saturation values of 1000 g/kg have to be postulated and which corresponds to a degree of adsorption which permits no structure in the polymer or accessibility factor. Thus the weight of probability is against the idea of the operation of a Langmuir isotherm.

Reference has been made already to the precondition for a site adsorption process to give a partition isotherm (Chapter 2.IIB) when appropriate relationships between interaction energies exist. The requirement is generally that the interaction energy between the molecules in the dye crystal and in the polymer should not differ greatly from that between adsorbed dye molecules and the polymer. This enables dye and polymer to be mixed without any great change in the potential energy of the system. It is possible to assess whether these conditions are met as a result of the work of Marjury (1954) and of Jones (1958) in which the two component (dye vapour–cellulose acetate) system was studied. The absence of a solvent in this study greatly simplifies the interpretation of the results as will be seen. Measurements were carried out to determine the isosteric heats of sublimation of disperse dyes and of adsorption on cellulose acetate. Both were found to be constant and of the same order of magnitude. This condition is exemplified by the data of Table 7.4. based on the work of Jones and Seddon (1964, 1965).

TABLE 7.4. A comparison of isosteric heats of adsorption and sublimation (Jones and Seddon, 1964, 1965).

Dye	ΔH_{iso} (kcal/mole) (*Heat of adsorption*)	• ΔH_{sub}^0 (kcal/mole) (*Heat of sublimation*)
Azobenzene	−17·8	+15·5
4-aminoazobenzene	−28·6	+26·5
1-methylamino anthraquinone	−21·8	+29·6
1-hydroxyethylamino anthraquinone	−37·3	+36·5

Adsorption of dye (1-methylamino anthraquinone) was found to have very little effect on the crystallinity of the film but the apparent second order transition temperature was lowered by 8°C. The extensibility of dyed film at any temperature was found to be greater for dyed than undyed film and the shrinkage temperature was reduced by 10°C. These observations indicate that the dye is capable of competing with cohesive forces within the film and consequently acting as a plasticizer. Other evidence for this effect has been provided by Cometto (1971) who has demonstrated the self "carrier" action some disperse dyes in polyethylene terephthalate.

The similarity of the enthalpies of dyeing from the vapour phase and of sublimation is also to be seen when the enthalpy of dyeing and solution are compared for aqueous dyeing systems. In Table 7.5 data are shown for dyeings on cellulose acetate from aqueous solutions taken from the work of Marjury (1954), Patterson and Sheldon (1960) and Bird and Harris (1957).

TABLE 7.5.   Enthalpy of dye absorption and solution in disperse dye/water/cellulose acetate system.

Dye	ΔH^0_{dyeing} (kcal/mole)	$\Delta H^0_{solution}$ (Kcal/mole)
Azobenzene	−11·80	11·20
4-aminoazobenzene	−10·40	9·20
4'-nitro-4-aminoazobenzene	−13·30	10·60
1-methylamino anthraquinone	−10·00	8·40
1-hydroxyethylamino anthraquinone	−13·50	11·30

The effects mentioned taken together provide a strong indication of the possibility of an ideal mixing situation giving a partition isotherm. It is noteworthy that in the aqueous system the enthalpy of adsorption and of solution are lower than those of adsorption from the vapour phase and of sublimation. An interesting comparison of the corresponding entropy changes has been made by Thompson (1969) and the values are shown in Table 7.6.

TABLE 7.6.   Entropies of dyeing from aqueous and vapour phases, of sublimation and of solution (Thompson, 1969).

Dye	Vapour phase		Aqueous phase	
	ΔS^0_{dyeing} (e.u.)	$\Delta S^0_{sublimation}$ (e.u.)	ΔS^0_{dyeing} (e.u.)	$\Delta S^0_{solution}$ (e.u.)
Azobenzene	36·3	19·6	26·4	19·8
4-aminoazobenzene	56·8	11·7	27·1	12·1
4'-nitho-4-aminoazobenzene	47·0	27·4	28·7	15·6
1-methylamino anthraquinone	49·9	35·8	30·4	15·6
1-hydroxethylamino anthraquinone	59·2	14·3	32·0	19·6

Clearly whether the dye is applied from the vapour phase or aqueous solution, it achieves a lower state of order in the adsorbed state. It appears also to be in a lower state of order when adsorbed from the vapour phase but this effect may be illusory because no allowance has been made for possible differences in the effect of dye-polymer interaction on the entropy of the polymer under the two sets of conditions, or the role of water in the dyeing process.

From Tables 7.4, 7.5 and 7.6 it is clear that water is not simply an inert transfer medium in the dyeing process acting simply as a system in which a molecular dispersion can be formed. The possibility that the solvent from which the dyeing took place plays a significant role in the dyeing of disperse dyes was first put forward by Heit *et al.* (1959) who examined dyeing behaviour with several solvents.

It was observed that although the disperse dye C.I. Disperse Red 1 give linear partition isotherms when applied from aqueous dyebaths (Bird and Harris, 1957), the same dye applied from other solvents in which it had sparing solubility showed curvilinear isotherms as shown in Fig. 7.6.

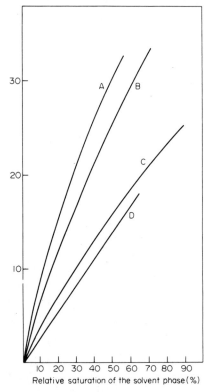

Fig. 7.6.   Adsorption isotherms for C.I. Disperse Red I from various solvents at 25°C. A, benzene; B, carbontetrachloride; C, methanol; D, butyl acetate. (Heit *et al.*, 1959).

In the isotherms the concentration of dye in solution as shown in Fig. 7.6 has been normalized, i.e. it has been expressed as a proportion of the saturation value in solution. This means that any given value of N/N_s corresponds to a single partial molar free energy in solution irrespective of the solvent. Since the uptake of dye by the fibre is a variable with the solvent at any N/N_s value then clearly the solvent is not simply an inert transfer medium but clearly involved in the dye–fibre interaction itself. The differences cannot be explained on the basis of fibre swelling since the solvents produce swelling effects in inverse order compared with their dyeing effect, i.e. benzene causes least swelling but most dye uptake, butyl acetate causes most swelling but least dye uptake. From this result and other more marginal data White (1960) concluded that a site adsorption mechanism operates in the uptake of disperse dyes by cellulose acetate and water molecules (or other solvents) play a competitive role. The competitive role of the water was regarded by White as capable of distorting the Langmuir isotherm which would obtain if water was not present. In support of his argument White showed that the isotherm equation which can be derived for two adsorbants, the solvent s and the dye d, competing for the same sites, namely

$$\theta d = \frac{N}{N_s}\left[\frac{1}{k_d} + \frac{k_s}{k_d} + \frac{N}{N_s}\left(1 - \frac{k_s}{k_d}.N_s\right)\right]^{-1} \tag{7.5}$$

gave a partition isotherm for low values of k_d, the adsorption coefficient of the dye and appropriate values of k_s/k_d where k_s is the adsorption coefficient of the solvent. The difficulty with this argument is that in order to get linearity either k_d has to be very much smaller than is observed in practice or the affinity of the solvent has to be very much higher than experimental data suggest it is. White's argument is consequently logically consistent but divorced from the situation which it purports to deal with. The argument falls down, moreover, because partition isotherms are observed when water is totally absent in vapour phase adsorption (Marjury, 1956). White's model gives a strongly curved isotherm when solvent adsorption is absent (i.e. $k_s = 0$).

Nevertheless it has been shown that White's assessment of the role of the dyebath solvent was qualitatively correct. Investigations by Thompson (1969) have shown that water plays a considerable role in the dyeing process beyond its action as a solvent. He showed also that there is a sufficient degree of non-reversibility in the dye–water–polymer system to make judgements based solely on the magnitude of thermodynamic parameters very much open to question. The competitive role of water was demonstrated by the effect of its presence on the maximum dye uptake on cellulose acetate. By carrying out dyeings in the vapour phase, Thompson was able to control the amount of water present so as to produce dyeings in the presence of varying amounts of water uptake. The results are illustrated in Fig. 7.7.

The relationship between maximum dye uptake and water adsorption was found to be exponential and fitted the empirical equation,

$$\text{Azo benzene uptake} = 10.72 \exp(-0.148 \times \text{water uptake}) \tag{7.6}$$

in which uptake values are expressed in weight % of dry cellulose acetate. The competitive effect of this kind indicates that the process is not one involving a simple displacement by competition for the same adsorption site as envisaged by White but a more complex phenomenon. This is indicated particularly by the fact that the number of molecules of water required to "displace" one molecule of azo benzene varies between 6 at low water uptake values to 20 at high water uptake values. This effect combined with the fact that in any case several water molecules are involved in each displacement suggests that the competition may be allosteric in nature. This means that the uptake of water by the polymer brings about configurational changes which destroy potential adsorption sites for the dye. Competitive adsorption of this kind is not always reversible because it is not necessarily true that the

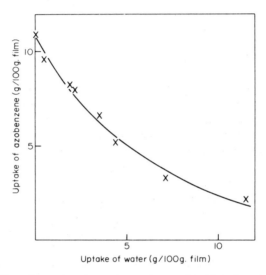

Fig. 7.7.   The effect of water uptake on the maximum adsorption of azo benzene by cellulose acetate at 95°C (Thompson, 1969).

displaced molecules can operate in the contrary sense. Certainly Thompson found that the competition between water and azo benzene was not truly reversible. This is shown by a hysteresis effect resulting in apparent equilibrium situations which depend upon how they are reached as is demonstrated in Table 7.7 which shows some data for cellulose acetate at 95°C.

It can be seen that there is little difference between the situations in which the two adsorbates are applied simultaneously and when water is adsorbed first. This is not surprising in view of the relatively rapid rate of diffusion of water. However when azo benzene is adsorbed first there is not only a reduction in the total adsorption, but the amount of water taken up at equilibrium is reduced. Thus azo benzene is able to interact with cellulose acetate in such a way that water cannot be adsorbed. The non-reversibility of the process is emphasized by the rates of equilibration. Exposure of a dyed

TABLE 7.7.   Comparative dye and water uptake values obtained by different routes to equilibrium (Thompson, 1969).

Method of "equilibration"	Total of dye and water (g.moles.kg^{-1})	Adsorption of azo benzene (g.moles.kg^{-1})	Adsorption of water (g.moles.kg^{-1})
Sorption of water and azo benzene vapour simultaneously	6·94	0·14	6·81
Sorption of azo benzene after equilibration with water vapour	6·51	0·12	6·39
Sorption of water vapour after equilibration with azo benzene vapour	4·20	0·26	3·94

film to water vapour results in a fairly rapid equilibration, but the rate at which dye molecules can displace water from a previously equilibrated film is much slower. The effect shown in Table 7.7 applies for one water vapour pressure (584 torr at 95°C) but it is evident at all values as is shown in Fig. 7.8. There is evidence also that adsorbed dye causes changes in the polymer whether water is present or not. The effect on adsorption is not large but is significant. It can be shown by an examination of maximum adsorption values from adsorption data obtained in dry dye vapours. The maximum adsorption value can be determined in two ways. Either the experimental isotherm can be extrapolated to the saturated vapour pressure value or the polymer may be exposed to a saturated vapour. Both results are obtainable from verifiable experimental situations but they are not the same. The discrepancy can be either positive or negative according to the dye used so that errors due to surface condensation etc. may be ruled out. Typical data are shown in Table 7.8.

TABLE 7.8.   Maximum adsorption values for different disperse dyes on cellulose acetate applied from the dry vapour phase (Thompson, 1969).

Dye	Temperature (°C)	Maximum adsorption values (g/100g dry film)	
		by extrapolation	by adsorption
Azo benzene	130	11·7	12·9
1-methylamino	145	10·1	8·9
	150	11·2	9·6
	155	11·3	9·3
1-hydroxyethylamino anthraquinone	164	14·4	19·4

Thus it is clear that the complexities of the dye-polymer system and the evident non-reversibility prevent any simple all embracing theoretical model

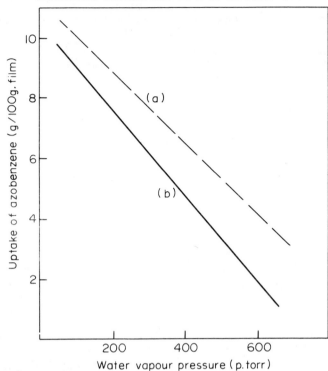

Fig. 7.8.   The uptake of azo benzene as a function of water vapour pressure on cellulose acetate at 95°C. (a) pre-equilibriated with azo benzene vapour (b) pre-equilibriated with water vapour. (Thompson, 1969).

of the dyeing process being developed. It is fairly clear in a qualitative sense that the dye molecules are specifically adsorbed on sites with binding energies comparable with those operating in the dye solids and the cohesive forces in polymers so as to give isothermal behaviour comparable with ideal mixing processes. The adsorption process is evidently such that some modification of the polymer inevitably occurs when extensive adsorption occurs and allosteric competition may be exhibited when solvents (including water) are present.

IV   INTERACTIONS BETWEEN DISPERSE DYES

One effect which is predictable on the basis of the ideal mixing type of behaviour of disperse dyes is the additivity of isotherms. In other words when a mixture of dyes is applied to a substrate they should not compete with one another despite the operation of a site mechanism. In this respect disperse dyes should behave differently from other dyes with which the neutral repulsion due to the ionic charge imposes a powerful limiting factor on

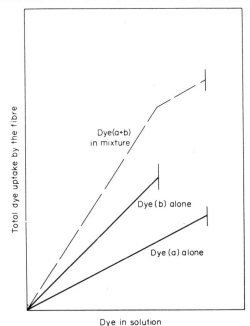

Fig. 7.9.   The additivity of isotherms with disperse dyes.

adsorption. The effect of the additivity of isotherms is shown schematically in Fig. 7.9.

It can be seen that the additivity can be put to very useful practical effect. By using a mixture of two dyes of very similar shade (perhaps a mixture of isomers) concentrations of dye in the fibre can be achieved which greatly exceed the maximum values obtainable with either dye resulting in the production of heavy shades. The use of appropriate mixtures of these dyes has been the subject of patent activity, (Ciba 1932, 1951). It has been claimed (Sandoz, 1957) that certain mixtures of dyes can give heavier shades than predicted by the addition of their individual adsorption values. No supporting evidence has been published although there is no theoretical reason why such an effect should not be achieved.

Not all disperse dyes show additive behaviour and it was shown by Vickerstaff and Waters (1942) that certain mixtures give strong evidence of interaction between dye molecules leading to lower than expected total adsorption values. A series of dyes based on in which R_1, was a methyl, β hydrothethyl, isobutyl and ethyl residue were examined. Interaction leading to non-additive isotherms was quite specific, being present when the ethyl substituted dye was mixed with the methyl or β-hydroxyethyl substituted dyes but in none of the other binary mixtures. Schuler and Remington (1954) showed interaction between 1-amino-4-hydroxy- and 1;4-dihydroxy anthraquinone. The early studies were extended by Bird and Rhyner (1961) and a very complex situation was revealed. Binary and some ternary mixtures of eight dyes were examined showing interaction in some cases. When ternary mixtures were used there were cases where the interaction of two of the components was apparently supressed. Typical results are shown in Table 7.9.

TABLE 7.9. Maximum adsorption values obtained using mixtures of disperse dyes on cellulose acetate at 85°C (Bird and Rhyner, 1961).

Mixture of dyes	Maximum adsorption values (mg/g)		Difference (%)
	Calculated	Observed	
I + II	10·2	10·3	+1
I + IV	7·5	4·7	−37
I + V	8·1	8·7	+7
I + VI	12·1	11·3	−7
I + VII	13·3	13·6	+2
V + VI	8·0	6·2	−23
V + VII	9·2	9·6	+4
VI + VII	13·2	9·6	−27
I + V + VI	14·1	12·5	−11
I + V + VII	15·3	15·5	+1
I + VI + VII	19·3	19·5	+1
V + VI + VII	15·2	11·4	−25

In Table 7.9 the numbers refer to dyes based on

where R_1, R_2, R_3, R_4 are:

Dye	R_1	R_2	R_3	R_4
I	CH_3	H	H	H
II	C_2H_4OH	H	H	H
IV	C_2H_5	H	H	H
V	CH_3	H	H	CH_3
VI	CH_3	CH_3	H	H
VII	CH_3	H	CH_3	H

The great specificity of the interactions can be clearly seen. The positive values which would appear to support the Sandoz (1957) claims are regarded by Bird and Rhyner as being due to experimental error but no evidence is produced in support of this view. However there are some indications that the effect may be real.

Bird and Rhyner determined the solubilities in water in the presence of dispersing agent of many of the dye mixtures used in the dyeing experiments and found that the interactions revealed by the unexpectedly low maximum adsorption values were exactly mirrored by the solubility data, i.e. the solubility of a mixture of dyes I and IV was 37% less than expected. However some of the positive values were also repeated in the solubility data so that the additive effect of mixing dyes may be as real an effect as the non-additivity.

Mixtures of constituents giving positive deviations from Raoult's Law are very common and consequently unexpectedly high solubility with appropriate mixtures of interactive dyes are to be expected. Certainly mixtures of isomers which have unexpectedly high solubility are known and generally arise from the surface activity of dyes. The work of Bird and Rhyner was carried out in the presence of a surfactant which was a derivative of sodium μ-heptadecyl benzimidazole disulphonate. Synergistic aqueous solubility effects in the presence of anionic surfactants have been reported in relation to oleophilic dyes when weakly polar or even non-polar additives were present. McBain and Green (1947), for example, have shown that the solubility of C.I. Pigment Orange 13 in potassium laurate and potassium oleate solutions is enhanced when benzene, toluene or hexane is present. It seems hardly possible that in a mixture of isomers the effective solubilities of *both* components can increase but that of the less soluble component could be increased depending on the nature of the interaction between the solute molecules and the surfactant. The effects which could be produced are not likely to be large and this probably explains why rather more work has been carried out on those cases where solubilities and maximum adsorption values are unexpectedly low.

Some of the cases of low dye uptake from mixtures observed by Bird and Rhyner and by Schuler and Remington have been examined closely by Johnson, Peters and Ramadan (1964) by a study of the solid state properties of dye mixtures. Non-interacting dyes giving the expected additivity of isotherms were found to show normal meeting point diagrams with appropriate eutectic points. Interacting dyes on the other hand showed no eutectic formation and consequently were capable of forming mixed crystals. Typical melting point diagrams are shown in Fig. 7.10.

Johnson *et al.*, concluded that interacting dyes formed mixed crystals in the polymer thus showing abnormal adsorption behaviour. This view has been disputed by Hoffman *et al.* (1968) who showed that there was no evidence for mixed crystal formation in the fibre while there was evidence of

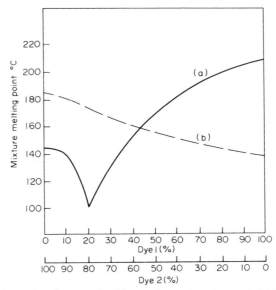

Fig. 7.10. Melting point diagrams for (a) non interacting dyes and (b) interacting dyes (Johnson *et al.*, 1964).

supressed aqueous solubility as observed by Bird and Rhyner. They concluded that mixed crystal formation was not significant and that interactions in solution were the main factor. This view is almost certainly the correct one. At the solubility limit of the dye mixture the equilibrium will be between the dye solution and the mixed crystals and it is the lower solubility of these which will give the lower maximum adsorption values.

Further examination of the problem by Daruwalla *et al.* (1970) has confirmed the effect of interaction in solution of interacting disperse dyes and has shown that the complex may be able to be adsorbed as such by the fibre.

V   REFERENCES

Biederman, W. (1971) *J. Soc. Dyers and Colourists*, **87**, 105.

Bird, C. L. (1954) *J. Soc. Dyers and Colourists*, **70**, 68.

Bird, C. L., Manchester, F. and Harris, P. (1954) *Discussions of the Faraday Soc.*, **16**, 85.

Bird, C. L. and Harris, P. (1957) *J. Soc. Dyers and Colourists*, **73**, 199.

Bird, C. L. and Firth, J. (1960) *J. Text. Inst.*, **51**, T1342.

Bird, C. L. and Rhyner, P. (1961) *J. Soc. Dyers and Colourists*, **77**, 12.

Blacker, J. G. and Patterson, D. (1969) *J. Soc. Dyers and Colourists*, **85**, 598.

Ciba, A. G. (1932) British Patent Specification 428767.

Ciba, A. G. (1951) British Patent Specification 654551.

Clavel, R. and Stanisz, T. (1923) *Rev. gén Matières col.*, **28**, 145–167.

Clavel, R. and Stanisz, T. (1924) *Rev. gén Matières col.*, **29**, 94, 122, 158.

Cometto, C. (1971) M.Sc. Thesis, Leeds University.

Daruwalla, E. H., Rao, S. S. and Tilak B. D. (1960) *J. Soc. Dyers and Colourists*, **76**, 418.

Daruwalla, E. H., Patel, R. M. and Tripathi, K. S. (1970) *Proc. 18th Hungarian Text. Conf.*, **2**, 173.

Datye, K. V., Pitkar, S. C. and Parao, U. M. (1971) *Textilveredlung*, **6**, 593.

Heit, C., Moncrieff-Yeates, M., Palin, A., Stevens, M., and White, H. J. (1959). *Text. Res. J.*, **29**, 6.

Hoffman, K., McDowell, W. and Weingarten, R. (1968) *J. Soc. Dyers and Colourists*, **84**, 306.

Husy, H. Merian, E. and Schetty, G. (1969) *Text. Res. J.*, **39**, 94.

Johnson, A., Peters, R. H., and Ramadan, A. S. (1964) *J. Soc. Dyers and Colourists*, **80**, 129.

Jones, F. (1958) Ph.D. Thesis, Leeds University.

Jones, F. and Flores, L. J. (1970) US Army Natick Laboratories Technical Report, 71-52-CE.

Jones, F. and Flores, L. J. (1971) *J. Soc. Dyers and Colourists*, **87**, 304.

Jones, F. and Seddon, R. (1964) *Text. Res. J.*, **34**, 373.

Jones, F. and Seddon, R. (1965) *Text. Res. J.*, **35**, 334.

Kartaschoff, V. (1925) *Helv. Chim. Acta*, **8**, 928.

Lauer, K. (1932) *Kolloid Z.*, **61**, 9.

Marjury, T. G. (1954) *J. Soc. Dyers and Colourists*, **70**, 442.

Marjury, T. G. (1956) *J. Soc. Dyers and Colourists*, **72**, 41.

Marsden, R. J. B and Urquhart, A. R. (1942) *J. Text. Inst.*, **33**, T105.

McBain, J. W. and Green, A. A. (1947) *J. Phys. Colloid Chem.*, **51**, 286.

Mechel, L. and Stauffer, H. (1941) *Helv. Chim. Acta*, **24**, 151.

Merian, E. (1966) *Text. Res. J.*, **36**, 612.

Millson, H. E. and Turl, L. H. (1951) *Text. Res. J.*, **21**, 685.

Patterson, D. (1954) *Discussions of the Faraday Soc.*, **16**, 247.

Patterson, D. and Sheldon, R. P. (1960) *J. Soc. Dyers and Colourists*, **76**, 178.

Sandoz, A. G. (1957) British Patent Specification 764439.

Schuler, M. J. and Remington, W. R. (1954) *Discussions of the Faraday Soc.*, **16**, 201.

Seddon, R. (1962) Ph.D. Thesis, Leeds University.

Stubbs, A. E. (1954) *J. Soc. Dyers and Colourists*, **70**, 120.

Thompson, J. (1969) Ph.D. Thesis, Leeds University.

Valkó, E. I. (1937) *Kolloid Chemische Grundlagen der Textilveredlung*, Springer, Berlin.

Vickerstaff, T. (1954) *The Physical Chemistry of Dyeing*, 2nd ed., Oliver and Boyd, London.

Vickerstaff, T. and Waters, E. (1942) *J. Soc. Dyers and Colourists*, **58**, 116.

White, H. J. (1960) *Text. Res. J.* **30**, 329.

Chapter 8

Fibre-reactive dyes

The fibre-reactive dyes constitute one of the largest and fastest growing ranges of dyes. For cellulosic materials they are probably the most important single range of dyes. Although they are of less importance in the field of colouration of man-made fibres this is largely due to the non-availibility of fibre-reactive dyes for the majority of such materials, viz. polyester, acrylic and polypropylene. This deficiency is not likely to be permanent in view of the technical advantages which would follow from its remedy.

The organic chemistry of reactive dyes is considered in great detail elsewhere (Beech, 1970) and will not be considered here except where special considerations demand. For convenience the physico-chemical aspects of fibre-reactive dyes will be considered in relation to cellulose on the one hand and other substrates on the other.

I FIBRE-REACTIVE DYES ON CELLULOSIC SUBSTRATES

The formation of coloured cellulose derivatives has been studied since the last decade of the nineteenth century. For various reasons the investigations did not lead to a commercially useful dyeing system. The historical aspects of the ultimate development of such a system in the years 1953–1956 are described elsewhere (Dawson *et al.*, 1960; Rattee, 1965) and provide interesting indications of how such developments actually take place. Attention will be confined in the present context to physical chemistry.

Cellulose consists essentially of poly-1,4 D-glucopyranose (I) and consequently be regarded as a polyhydric alcohol undergoing most of the general reactions of alcohols.

Due to the electronegativity of oxygen atoms, hydroxyl groups are subject to ionization particularly in the presence of proton acceptors. This property is manifest, of course, in the simplest case of water. In alcohols the ionization behaviour is conditioned by the inductive effect of the alkyl or aryl substituents. In the case of methanol, for example, the release of electrons by the methyl group counteracts the inductive effect of the oxygen so that the alcohol is less acidic than water.

$$\text{General case} \qquad R - O \leftarrow H$$
$$\text{Methanol} \qquad CH_3 \rightarrow O \leftarrow H$$

In polyhydric alcohols the situation is complicated by near neighbour effects and also competitive effects. The overall result is, in general, to make polyhydric alcohols more acidic than simple alcohols and indeed comparable with water. This is shown by the dissociation constants (K).

$$
\begin{aligned}
&\text{Water} &&K = 2 \cdot 09 \times 10^{-14} \\
&\text{Methanol} &&K = 8 \cdot 1 \ \times 10^{-15} \\
&\text{Mannitol} &&K = 7 \cdot 5 \ \times 10^{-14} &&\text{(Souchey } et\ al., 1950) \\
&\text{Cellulose} &&K = 1 \cdot 84 \times 10^{-14} &&\text{(Neale, 1929).}
\end{aligned}
$$

The figure given above for $K_{\text{cellulose}}$ refers to the ionization of the hydroxyl group in the 6-position in the glucopyranose ring. The hydroxyl groups in the 2 and 3 positions are less acidic so that acidity decreases in the order 6-OH > 2-OH > 3-OH. Cellulose is consequently ionized under alkaline conditions and can act as a nucleophilic reagent in Schotten–Baumann reactions with compounds containing electron-deficient carbon atoms or unsaturated trigonal aliphatic carbon atoms. The relative reactivities of the hydroxyl groups in the different positions in the glucopyranose ring are not known. Studies by Gardner and Purves (1942) using toluene sulphonyl chloride as a substituting reagent produced relative reaction rates but no allowance was made for relative ionization of the groups in the different positions. This is an important factor as will be shown later.

A   The physical chemistry of dye-fibre reaction

All available fibre-reactive dyes for cellulose utilize the reactivity of ionized cellulose as a nucleophilic reagent and attention will be confined to the kind of reaction involved in such cases. Since the ionization of cellulose involves the presence of water, competitive reaction with hydroxyl ions must be considered at the same time as the desired alcoholysis reaction. Considering first of all a simple homogeneous system in which a reactive dye D, bearing an unspecified reactive system with an alcohol AOH in the presence of water, the following reactions occur.

$$(1)\quad \text{AOH} \rightleftharpoons \quad \text{AO}^- + \text{H}^+ \qquad \text{(ionization of the alcohol)}$$
$$(2)\quad \text{AO}^- + D \rightarrow \text{AO}D \qquad \text{(alcohysis)}$$
$$(3)\quad \text{OH}^- + D \rightarrow \text{HO}D \qquad \text{(hydrolysis)}$$

Considering now the reaction kinetics, the respective rates of alcoholysis and hydrolysis (df/dt and dh/dt) are given by

$$df/dt = k_f[D]_t[\text{AO}^-]_t \qquad (8.1)$$

$$dh/dt = k_h[D]_t[\text{OH}^-]_t \qquad (8.2)$$

where k_f and k_h are the bimolecular rate constants of fixation and hydrolysis respectively and the suffix t signifies concentrations at time t.

In most systems studied experimentally it is convenient to adjust conditions so that a large excess of alcohol is present in a buffered system so that $[\text{AO}^-]$ and $[\text{OH}^-]$ are constants. Under these circumstances both reactions follow first order kinetics, and the rate of disappearance of dye, (dD/dt) is given by

$$-dD/dt = df/dt + dh/dt = \{k_f[\text{AO}^-] + k_h[\text{OH}^-]\}[D]_t \qquad (8.3)$$

so that

$$k_{\text{tot}} = k_f^1 + k_h^1 = k_f[\text{AO}^-] + k_h[\text{OH}^-]$$
$$= \frac{1}{t}\ln\frac{[D]_0}{[D]_t} \qquad (8.4)$$

in which k_f^1 and k_h^1 are the first order constants for alcoholysis and hydrolysis respectively.

However, because of the stoichometric equation (1) the values of $[AO^-]$ and $[OH^-]$ are related. The ionization of the alcohol is governed by a dissociation constant K_a given by

$$K_a = \frac{[AO^-][H^+]}{[AOH]} \tag{8.5}$$

or

$$K_a = \frac{[AO^-]K_w}{[AOH][OH^-]} \tag{8.6}$$

where K_w is the dissociation constant of water.

Since alcohols are very weak acids, the degree of ionization is very small at any pH below 11 so that with the limiting condition $[AOH]$ is virtually independent of $[AO^-]$ and $[H^+]$. Hence

$$\frac{[AO^-]}{[OH^-]} = \frac{K_a[AOH]}{K_w} . \tag{8.7}$$

where $[AOH]$ is the gross concentration of the alcohol. Substituting equation (8.7) in equation (8.4) gives

$$k_{tot} = k_f \cdot \frac{K_a[AOH]}{Kw}[OH^-] + k_h[OH^-] = \frac{1}{t}\ln\frac{[D_0]}{[D_t]} \tag{8.8}$$

The reaction behaves consequently as a first order reaction in which the overall rate constant is dependent upon the hydroxyl ion concentration, i.e.

$$\frac{k_{tot}}{[OH^-]} = k_f \cdot \frac{K_a[AOH]}{Kw} + k_h = \text{a constant} \tag{8.9}$$

so that

$$\frac{k_f}{k_h} = \frac{Kw}{K_a[AOH]}\left[\frac{k_{tot}}{k_h[OH^-]} - 1\right] = R' \tag{8.10}$$

where R' is the reactivity ratio. Equation (8.10) is slightly different from that given elsewhere but equivalent. Thus the ratio of the bimolecular rate constants will be constant at any temperature, with a given alcohol at a constant concentration for a given dye. Equation (8.10) predicts that it is independent of the pH conditions under which the reaction occurs, and this has been confirmed experimentally (Fern and Preston, 1961; Ingamells et al., 1962). Under conditions where the degree of ionization of the alcohol is significant, $[AOH]$ becomes pH dependent and so does the reactivity ratio.

The efficiency of the alcoholysis reaction is governed by the ratio of the competing alcoholysis and hydrolysis reactions, (df/dh) and this is given by combining equations (8.1) and (8.2),

$$\frac{df}{dh} = \frac{k_f[AO^-]}{k_h[OH^-]} \tag{8.11}$$

Given the limiting condition of equation (8.7) and equation (8.10) this gives

$$\frac{df}{dh} = \frac{R'.K_a}{Kw}[AOH] \tag{8.12}$$

so that efficiency of reaction may also be constant under appropriate conditions in a homogeneous system. The value of K_a is important, however, in comparing different alcohols; a factor which was neglected in the work of Dawson et al., (1960) in comparing reactions with n- and isopropanol.

Attention may now be turned to the real technical situation in which the dyes are reacted with a water swollen cellulosic fibre in a two phase system. This is a much more complicated system since diffusion and adsorption as well as surface charge effects enter into the consideration. With regard to the external aqueous phase and the internal aqueous phase the situation will be covered by equation (8.1). This may be rewritten in the forms,

$$\left[\frac{dh}{dt}\right]_s = k_h[OH^-]_t^s[D]_t^s \tag{8.13}$$

$$\left[\frac{dh}{dt}\right]_i = k_h[OH^-]_t^i[D]_t^i \tag{8.14}$$

in which the indices s and i signify external and internal aqueous phases respectively.

The problem of the diffusion of a substance into a phase with which it can react according to first order kinetics has been considered by Danckwerts (1950) whose solution for an infinite plane slab and an infinite reagent solution (e.g. dyebath) may be written as

$$Q_t = [D]^F\left[\frac{\bar{D}_1}{k_f}\right]^{0.5}[tk_f^1 + 0.5]\,\text{erf}\,(tk_f^1)^{0.5} + \left[\frac{tk_f^1}{\Pi}\right]^{0.5}\exp\,(-tk_f^1) \tag{8.15}$$

in which Q_t is the amount reacted in time t, $[D]^F$ is the constant surface adsorbed concentration of the reagent, $\bar{D}$ is the diffusion coefficient (assumed concentration and distance independent), k_f^1 is the first order reaction constant. Sumner and Weston (1963) have shown that under appropriate limiting conditions equation (8.15) may be simplified to

$$Q_t = [D]^F\left[t + \frac{1}{2k_f^1}\right](\bar{D}\,k_f^1)^{0.5} \tag{8.16}$$

Taking the simpler form of the equation the rate of fixation is given by

$$dQ/dt = [D]^F(\bar{D}\,k_f^1)^{0.5} \tag{8.17}$$

The total rate of hydrolysis is made up of that in the external solution and the internal aqueous phase. If all the dye adsorbed by the film could be chemically fixed, the efficiency of fixation E would depend upon the distribution of the dye. This would give rise to the ratio of the two reaction rates from equations (8.13) and (8.17) i.e.

$$E = \frac{dQ}{dh} = \frac{[D]^F (\bar{D} k_f^1)^{0.5}}{[D]^s k_h [OH]^s} \tag{8.18}$$

However, such a level of efficiency can never be achieved due to hydrolysis of dye in the film. The true value of E will be given by

$$E = \frac{[D]^F [\bar{D} k_f^1]^{0.5}}{k_h [OH]^s [D]^s + k_h [OH]^i [D]^i} \tag{8.19}$$

The values of $[OH]^s$ and $[OH]^i$, $[D]^s$ and $[D]^i$ are related since they are concentrations modified by a surface potential. The presence of carboxyl groups and the ionization of the cellulose will give rise to a potential ψ_{cell}. This being negative, and assuming the dye to bear a single negative ionic charge,

$$\frac{[D]^s}{[D]^i} = \frac{[OH]^s}{[OH]^i} = \exp - \frac{e\psi}{kT} \tag{8.20}$$

Combining equations (8.19) and (8.20) gives

$$E = \frac{[D]^F \{\bar{D} k_f^1\}^{0.5}}{k_h [OH]^s [D]^s (1 + \exp 2e\psi/kT)} \tag{8.21}$$

Equation (8.21) may be expressed as

$$E = \frac{[D]^F}{[D]^s} \left[\frac{\bar{D}}{[OH]^s} \cdot \frac{k_f^1}{k_h^2} \cdot \frac{1}{[OH]^s (1 + \exp 2e\psi/kt)} \right]^{0.5} \tag{8.22}$$

The term k_f^1 may be expressed in terms of a bimolecular rate constant as $k_f [\text{cello}^-]$ in which $[\text{cello}^-]$ represents the ionized cellulose concentration. $[\text{Cello}^-]$ is related to $[OH]^i$ by a relationship analogous to equation (8.6) so that,

$$\frac{[\text{cello}^-]}{[OH]^i} = \frac{K_{cell} [\bar{C}]}{Kw} = K_{cell}^1 \tag{8.23}$$

in which $[\bar{C}]$ is the cellulose concentration.

Combining equation (8.22) with (8.23), (8.20) and (8.10) gives

$$E = \frac{[D]^F}{[D]^s} \left[\frac{\bar{D} . R^1_{cell} . K^1_{cell}}{k_h [OH]^s} \cdot \frac{\exp e\psi/kT}{(1 + \exp 2e\psi/kt)^2} \right]^{0.5} \tag{8.24}$$

Providing the surface potential is more negative than $-14\,\text{mV}$ the exponential term decreases with ψ. This condition will apply with cellulose under any likely fixation conditions. So as far as experimental variables are concerned the efficiency of fixation will fall as the substantivity ratio and the surface potential fall and as the pH or reactivity of the dye increase. The efficiency will be in fact sensitive to the pH for two reasons not explicit in equation (8.24). Firstly the hydroxyl ion is potential determining in that as the pH increases, the cellulose becomes progressively ionized so that ψ becomes more negative. In addition the electrical potential affects the affinity of the dye for the cellulose (see equation (6.11) so that the substantivity ratio falls as the cellulose becomes more ionized. From a general knowledge of the characteristics of dye adsorption isotherms for cellulose it is clear that the substantivity ratio $[D]^F/[D]^S$ will fall as the depth of shade increases, and since the dye anions bear a negative charge the depth of shade will modify the potential term again unfavourably to high fixation efficiency. Equation (8.24) has been derived for a system with a particular geometry (an infinite plane slab) while in practical terms interest lies in fibres and finite films. The change of the geometry of the adsorption/fixation system will, of course, modify equation (8.24) but in a qualitative sense the factors of importance will remain and the equation may be used as a consequence for a wide range of real systems in a general predictive sense. The mathematical solution for a cylindrical fibre is possible using Danckwert's (1951) solution to the diffusion equation

$$Q_t = 4\pi \bar{D} [D]^F \sum_{n=1}^{\infty} \frac{k_f^1 + \bar{D}\alpha_n^2 \exp\{-t(k_f^1 + \bar{D}\alpha_n^2)\}}{(k_f^1 + \bar{D}\alpha_n^2)} \tag{8.25}$$

in which the symbols have the same significance as before, and α_n are the successive roots of the Bessel function $J_0(r\alpha) = 0$. For accurate numerical predictions for cylindrical fibre systems computation using equation (8.25) combined with those discussed earlier would be necessary. However, for general purposes equation (8.24) may be regarded as giving a satisfactory representation of the situation.

The predictions of equation (8.24) have all been substantiated experimentally by Sumner and Taylor (1967) and in practice. Beckmann and

TABLE 8.1. Fixation efficiency values as a function of pH, temperature and shade depth calculated from the data of Beckmann and Hildebrand (1965).

Depth of shade	Temperature	pH$_s$	E
4%	40°C	10·0	1·020
2%			2·311
1%			2·846
2%	24°C	10·0	5·098
	40°C		2·311
	60°C		1·247
2%	40°C	10·0	2·311
		10·5	2·135
		11·0	2·077

Hildebrand (1965) have published data regarding the effect of temperature, pH and the depth of shade on the fixation of a dichloro quinoxalin-6-yl fibre reactive dye on cellulose from which the following efficiency values shown in Table 8.1 have been calculated.

Sumner (1965) has presented data showing the effect of pH on the substantivity ratio. The effect in a particular case is shown in Fig. 8.1. It can be seen that at pH values above 10·5 the effect can be very large.

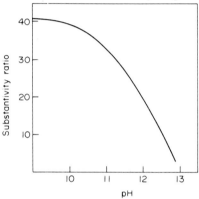

Fig. 8.1.   The effect of pH$_s$ on substantivity ratio.

In the discussion thus far the bimolecular rate constants of fixation and hydrolysis have been assumed to be constant. In fact with important ranges of fibre reactive dyes for cellulose this is not so. There are the dyes based on halogeno triazinyl or diazinyl reactive systems attached to the chromogen via an imino bridging group. e.g.

in which X, Y, Z are a variety of substituents including chlorine. The effect was first noted by Horrabin (1963) with model compounds and has been reported by several investigators who have examined the variability in some detail (Rattee, 1969; Ingamells et al., 1961; Sumner and Weston, 1963; Aspland and Johnson, (1965)). The reactive dyes of the triazinyl- or diazinyl-amino series show log k_h^1 vs pH relationships of three general kinds illustrated qualitatively in Fig. 8.2.

Dyes giving rectilinear relationships of type (a) bear an alkyl or aryl substituent on the bridging nitrogen atom and behave in the expected manner. Dyes showing a slight inflexion are found among those in which the triazinyl or diazinyl ring is attached to a derivative of H-acid (1-hydroxy 8 amino 3:6 disulpho naphthalene). There may be other sources of dyes of type (b) but they have not been reported. The majority of technical dyes of the type under consideration fall into category (c).

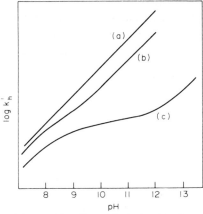

Fig. 8.2.   The three general forms of log k_h^1 vs pH relationship observed with tri- and diazinyl reactive dyes.

Horrabin (1963) proposed that the effect was due to acidity of the imino bridging group so that as the pH was increased it ionized to produce a dye molecule of reduced reactivity.

It is not at all easy to demonstrate whether Horrabin's explanation is correct since the apparent *pk* value for the ionization is about 10·5 for the chlorotriazines and 12·5 for the chlorodiazines. Thus direct titration particularly under conditions where hydrolysis of the reactive chlorine atom occurs, is difficult to carry out. It is not necessary, however, to assume any particular mechanism for the formation of two forms of the dye in order to derive the expected kinetic equations. It is only necessary to assume that two forms of the dye are in equilibrium in the presence of alkali in accordance with a stoichometric equation $A + OH^- \rightleftharpoons B$ which allows a kinetic analysis.

For the equilibrium

$$\frac{[B]}{[A][OH^-]} = K_d \tag{8.26}$$

However, the total amount of dye present is D given by

$$[A] + [B] = [D] \tag{8.27}$$

so combining (8.26) with (8.27) gives

$$[A] = \frac{[D]}{1 + K_d[OH]} ; [B] = \frac{K_d[OH][D]}{1 + K_d[OH]} \tag{8.28}$$

For the hydrolysis reactions in a buffered system,

$$-dA/dt = k_1[OH^-][A]_t \tag{8.29}$$

$$-dB/dt = k_2[OH][B]_t \tag{8.30}$$

Combining equations (8.24), (8.29) and (8.30) to give the total rate of hydrolysis,

$$-dD/dt = -\{dA/dt + dB/dt\}$$
$$= [OH]\{(k_1 + k_2[OH]K_d)/(1 + K_d[OH])\}[D]_t \tag{8.31}$$

Thus the kinetics of the overall hydrolysis will be first order with a bimolecular rate constant k_h given by

$$k_h = \frac{k_1 + k_2[OH]K_d}{1 + K_d[OH]} \tag{8.32}$$

Hence the bimolecular rate constant is pH dependent.
Equation (8.32) may be transformed to give

$$\frac{1}{k_h - k_2} = \frac{K_d}{(k_1 - k_2)} \cdot [OH] + \frac{1}{(k_1 - k_2)} \tag{8.33}$$

or

$$\frac{1}{1 + K_d[OH]} = \frac{1}{(k_1 - k_2)} \cdot k_h - \frac{k_2}{k_1 - k_2} \tag{8.34}$$

Using experimental data both equations (8.33) and (8.34) may be solved numerically to obtain values for k_1, k_2 and K_d. This has been done by Murthy and Rattee (1969) who examined the hydrolysis behaviour of three dichlorotriazinyl amino dyes of category (c) over a range of temperatures, and also one dye of category (b) at a single temperature. The dyes were indicated as follows,

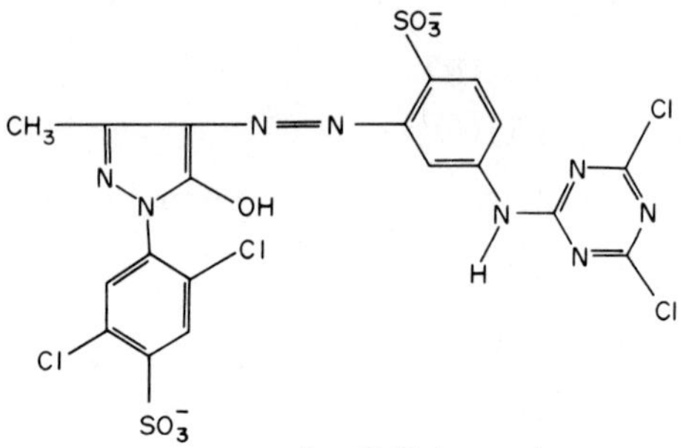

Dye II (Category c)

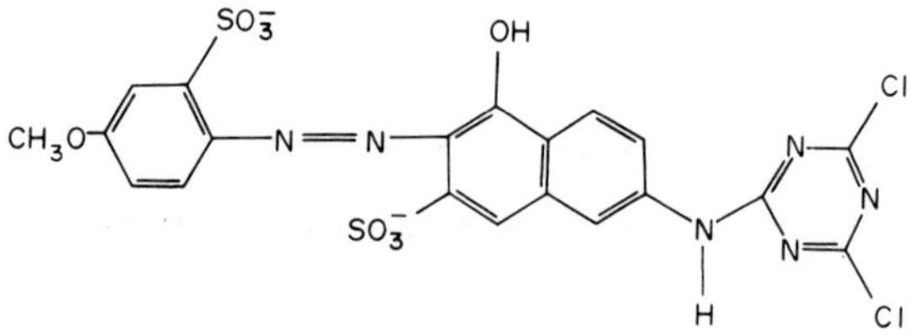

Dye III (Category c)

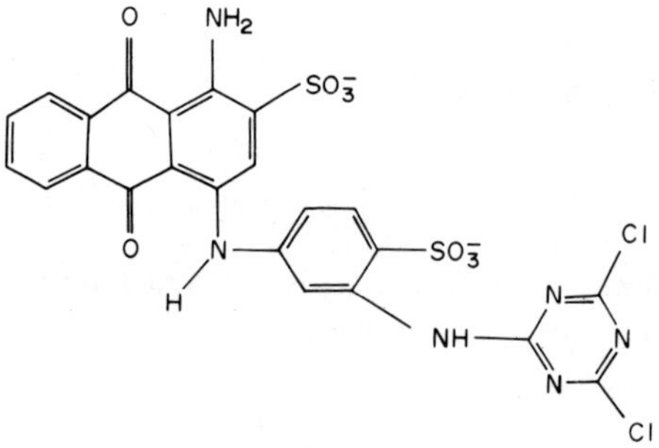

Dye IV (Category c)

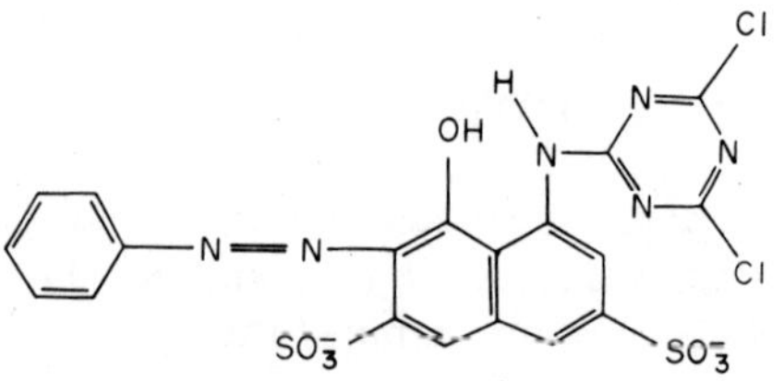

Dye V (Category b)

It was found that in the case of dyes II–IV the values of k_2 required to linearize the relationship of equation (8.33) were very small and within experimental error. In other words $1/k_h$ was proportional to [OH] over a range of temperatures giving a relationship of high probability. Two typical cases are illustrated in Fig. 8.3.

With dye V this did not apply. Computation of k_2 showed it to be comparable with k_1. Values for all four cases are shown in Table 8.2.

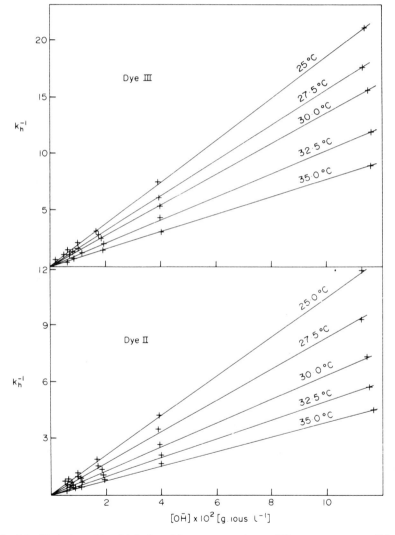

Fig. 8.3. Variation of k_h with hydroxyl ion concentration at different temperatures (Murthy and Rattee, 1969).

TABLE 8.2. k_1 and k_2 values for different dichlorotriazinyl dyes (Murthy and Rattee, 1969).

	Dye II	Dye III	Dye IV	Dye V
k_1 (1.mole.$^{-1}$ min^{-1})	9·60	5·76	14·28	12·70
k_2 (1.mole.$^{-1}$ min^{-1})	0	0	0	2·70
Temp. (°C)	27·5	27·5	28·0	27·5

Aspland and Johnson (1965) carried out similar experiments using dichloropyrimidinyl compounds. However, they considered their results in terms of an erroneous mathematical analysis which predicted $^1/k_h$ proportional to $^1/OH$. Not surprisingly this was found inapplicable. The error arises in the assumption that certain limiting conditions applied to their experimental situation when this was not so. Hughes (1970) carried out a study of two dichloropyrimidinyl compounds based on H-acid using the same approach as that employed by Murthy and Rattee. The two dyes had the following structures.

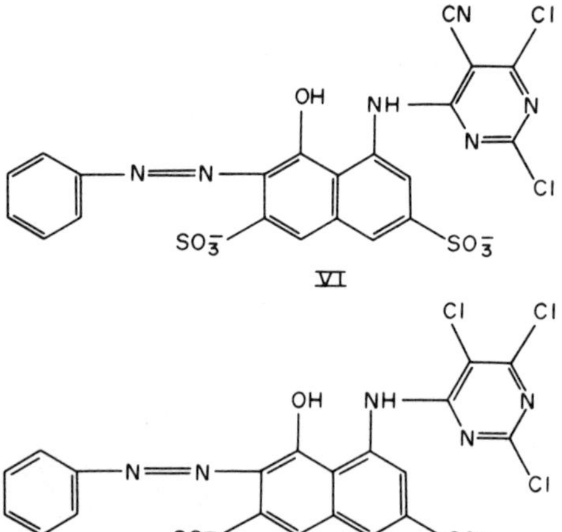

VI

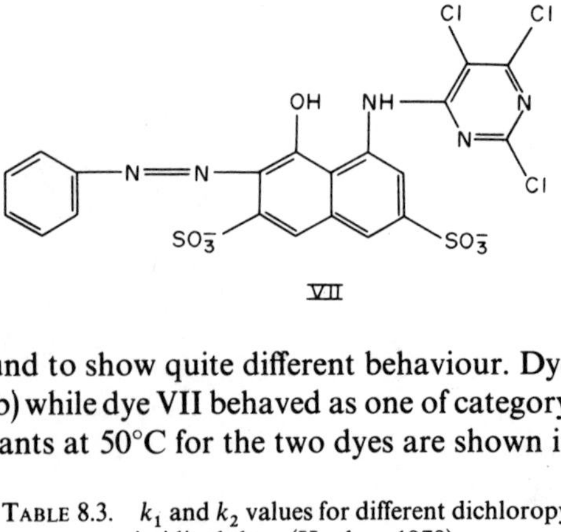

VII

They were found to show quite different behaviour. Dye VI behaved as a dye of category (b) while dye VII behaved as one of category (b). Second order hydrolysis constants at 50°C for the two dyes are shown in Table 8.3.

TABLE 8.3. k_1 and k_2 values for different dichloropyrimidinyl dyes (Hughes, 1970).

	Dye VI	Dye VII
k_1 (1.mole.$^{-1}$ min.$^{-1}$)	39·22	7·52
k_2 (1.mole.$^{-1}$ min.$^{-1}$)	0·210	0

The data of Murthy and Rattee and also of Hughes may be used to calculate a value of K_d. Typical values are shown in Table 8.4. The value of $-\log K_w/K_d$ is also shown. On the basis of Horrabin's theory this should be pK_a for the loss of a proton by the imino bridging group.

TABLE 8.4. K_d values for different chlorotriazinyl and diazinyl reactive dyes.

Dye	Temperature (°C)	K_d	$-\log K_w/K_d$ (pK_a)
II	27·5	$7·937 \times 10^2$	10·80
III	27·5	$8·865 \times 10^2$	10·79
IV	28·0	$7·855 \times 10^2$	10·77
V	27·5	$8·403 \times 10^2$	11·00
VI	50·0	$2·632 \times 10^2$	10·84
VII	50·0	$2·174 \times 10^1$	11·92

However the data of Murthy and Rattee does not support Horrabin's theory of ionization of the imino bridging group and the use of a pK_a term is not permissible. The values of k_1 and K_d were taken over a range of temperatures and it was found that the Arrhenius plot ($\log k_1$ against $1/T$) was not linear as shown in Fig. 8.4.

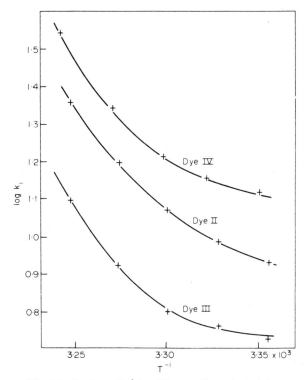

Fig. 8.4. $\log k_1$ vs T^{-1} for different chlorotriazinyl dyes.

This indicates that in the temperature range studied, activity changes or changes of state occur which lead to a variation in either the activity coefficient or the entropic factor. These results have been confirmed by direct calorimetry (Iyer and Jayaram, 1970). The behaviour of K_d with temperature was also unexpected. If the equilibrium described by K_d is an acid-base equilibrium as proposed by Horrabin, then it should behave as a typical weak acid ionization and exhibit a maximum value of K_d with temperature. The relationship between K_d and temperature for dyes II, III and IV is shown in Fig. 8.5.

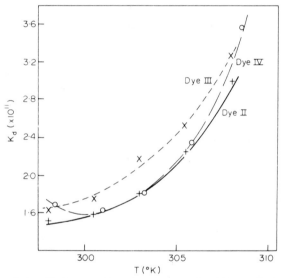

Fig. 8.5.   The relationship between K_d and temperature for different dichlorotriazinyl dyes.

The values of K_a clearly pass through a *minimum* value rather than a maximum as normally observed (Harned and Owen, 1958). It is possible that the way in which the apparent acid dissociation constant varies with temperature is connected with the activity changes with temperature revealed by the Arrhenius plot. Murthy was quite unable to detect any ionization of the imino bridging group by direct titration in the cases of Dyes II–IV and a large number of other instances (Murthy, 1967). Taken together with the kinetic data this suggests that Horrabin's hypothesis is not correct at least in relation to the dyes examined. A similar conclusion was reached by Rhys and Zollinger (1966) as a consequence of their study of hydrolysis. They suggested that the equilibrium involved in an isomerization rather than an ionization, see p. 259. Thus the matter of the variation of the hydrolysis constants with pH with reactive dyes in category (b) or (c) remains incompletely understood.

i.e.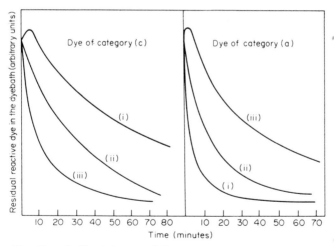

The variation does lead to important practical consequences in connection with the rate of fixation of the dyes. The rate of fixation to the fibre is governed by equation (8.17). It has already been shown that increasing the pH under conditions of constant ionic strength and temperature must lead to a fall in $[D]^F$ due to the electrostatic effect. Thus if $pH_1 > pH_2$, $[D]^F_{1 \text{ and } 2}$ and consequently $[D]^s_{1 \text{ and } 2}$. In addition to this the reaction rate does not increase very much with pH if the dye falls into category (c) as do most of those available in practice. Raising the dyebath pH is consequently equivalent to a desorption of dye from a phase in which it reacts rapidly due to its high concentration, into a dilute phase in which it will react more slowly and at the same time not compensating for this by increasing the effective reactivity. Thus raising the dyebath pH in such cases leads to a *reduction* in the rate of fixation rather than the increase which "common sense" might predict. This has been shown by Rattee (1963) and the effect is well illustrated by the rate curves of Fig. 8.6 which show the opposite effect of pH on the fixation rate of dyes in category (c) and category (a).

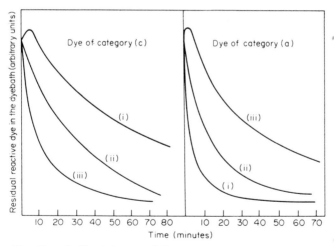

Fig. 8.6. The effect of pH on the rate of disappearance of reactive dyes in the presence of cellulose (Rattee, 1963).

The illustrations above show the rate of disappearance of reactive dye from a dyebath in two cases, C.I. Reactive Blue 1 and C.I. Reactive Red 5.

The alkaline conditions are indicated by (i) pH 12·5, (ii) pH 11·0 and (iii) pH 10·4 respectively. It can be seen that the dyes behave in opposite senses. C.I. Reactive Red 5 which is a category (a) dye behaves in an expected manner but C.I. Reactive Blue 1 which is a category (c) dye fixes more and more slowly as the pH is raised.

The possibility of two reactive forms of the dye in application under alkaline conditions is not confined to heterocyclic reactive dye systems. The so-called sulphonyl reactive dyes, i.e. the Remazol dyes of Farbwerke Hoechst, are employed in the first instance as the sulphato ethyl sulphonyl dyes and are converted to the vinyl form by mild alkaline treatment. Thus in the compound

$$R - SO_2 - CH_2 - CH_2 - O\,SO_3\,Na$$

VIII

the hydrogen atoms of the α-carbon are activated through electron withdrawal by the sulphonyl group. The ester group is a good leaving group so that formation of the vinyl sulphone proceeds readily.

$$R - SO_2 - CH_2 - CH_2 - OSO_3^- \xrightarrow{OH^-} R - SO_2 - CH = CH_2 + HSO_4^-$$

Due to the electrophilic nature of the sulphonyl residue direct nucleophilic attack is always possible to give the hydroxyethyl sulphonyl compound directly. This is a well known feature of such reactions and the relative importance of the two reactions is affected by concentration, temperature and substitution factors. In considering this Stamm (1964) regards the formation of the vinyl compound as decisive so that subsequent nucleophilic addition occurs through the vinyl sulphone step.

$$CH_2 = CH - SO_2 - \frac{+RO^-}{-RO^-} RO - CH_2 - CH - SO_2 - \frac{+H^+}{-H^+} RO - CH_2 - CH_2 - SO_2 -$$

However, there is evidence that the direct reaction may be important. If instead of the sulphate ester of the hydroxyethyl sulphone, the methane sulphonyl ester is used, hydrolysis of dye during application to cellulose is reduced (Beech 1970). Since the formation of the vinyl sulphone is rapid, this suggests that the competive direct hydrolysis is a significant factor. Direct evidence of this is not easily obtained. Investigations by Bhagwath and co-workers (1970) attempted to demonstrate that the sulphate ester form of the dye was actually the main reactive species present but it is difficult to produce a fully self consistent argument from their data without making some major assumptions. It was found that the rate of hydrolysis and the rate of fixation of the dye C.I. Reactive Blue 19 was more rapid when it was used in the sulphate ester form than when in the vinyl sulphonyl form. This is illustrated in Fig. 8.7.

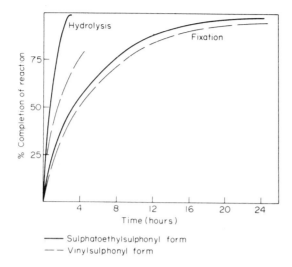

Fig. 8.7.   Hydrolysis and Fixation of different forms of C.I. Reactive Blue 19.

The data suggest that the sulphate ester form of the dye hydrolyses and fixes to the fibre more rapidly than does the vinyl sulphonyl form. However, there is a contradiction in that these data are not consistent with the fact that under the conditions of hydrolysis (pH ~ 12) conversion of the sulphate ester to the vinyl sulphone is very rapid, and under the conditions of fixation pH 10·5 it takes no more than 5–10 minutes. If conversion occurs so rapidly then the different curves must be referring to events after conversion and do not support the hypothesis advanced. The hydrolysis/conversion ratio may be affected by pH and temperature but the fixation data requires a more complex explanation for which inadequate data are provided.

B   The physical chemistry of dye fastness

In this context fastness is taken to mean fastness to wet treatments. In relation to dyes of other, non-reactive, classes the fastness to wet treatments relates to the degree and rate of desorption of the dye under the conditions in question. However, with reactive dyes there is a slightly more complex situation. Firstly there is present at the end of the dyeing process due to competitive hydrolysis some unfixed dye on the fibre. This dye has low affinity and unless it is removed the dyeing will have a low fastness to water or perspiration. Thus the removal of unfixed dye at the end of the dyeing process is an important stage in the operation. When this has been achieved, all the dye present on the fibre is covalently bound and the subsequent fastness depends entirely on the chemical stability of the dye–fibre bond. This provides the second aspect of dye fastness with reactive dyes.

1   *The removal of unfixed dye*

Due to the fact that the wet fastness of fibre-reactive dyes derives from the covalent binding, the substantivity of the dye needs to be no more than that needed to give satisfactory fixation efficiency and rate. Consequently removal of unfixed dye under conditions of low ionic strength and high temperature would be expected normally to be a rapid and simple process. This is in fact the case, but there are circumstances in which charge effects due to the sulphonate groups in the fixed dye can slow down dye removal. These are more interesting from the theoretical than the practical point of view, but this need not always be the case.

At the end of the dyeing and fixation process the unfixed dye anions in the cellulose exist in an environment of fixed dye anions, i.e. in an atmosphere of sulphonated groups. This is a highly unfavourable situation for adsorption and when the salt is removed from the fabric diffusion out of the cellulose should be rapid. However, if the fixation is not uniform, e.g. if the dye is fixed largely near the fibre surface, then the dye must diffuse into and through a zone of high negative potential. This is shown diagramatically in Fig. 8.8. Certain application methods with highly reactive dyes can give rise to ring

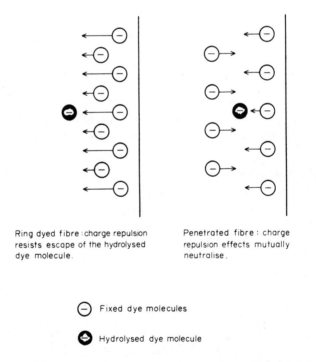

Ring dyed fibre : charge repulsion
resists escape of the hydrolysed
dye molecule.

Penetrated fibre : charge
repulsion effects mutually
neutralise.

⊖ Fixed dye molecules

❸ Hydrolysed dye molecule

Fig. 8.8.   A schematic representation of the effect of ring dyeing on the removal of unfixed dye.

dyeing, and when this occurs unusually slow removal of unfixed dye is observed. In batch application this may not be so serious as in continuous production which tends to be based on fixed times of sub-processing. The addition of electrolyte to the washing liquor aids removal of dye when this problem arises, although normally low ionic strengths are desirable.

2   The stability of the dye–fibre bond

It is well known in general cellulose chemistry that classical esters, e.g. cellulose nitrate, are readily hydrolysed while classical ethers, e.g. methyl cellulose, are not. The two kinds of compound are correspondingly easy and difficult to prepare respectively. Satisfactory fibre-reactive dyes are required to be fixed fairly readily and to give reasonably stable bonds. Accordingly, they do not fall into the simple classifications of standard chemistry. The compounds formed by reaction between cellulose and fibre-reactive dyes may be regarded as activated ethers or de-activated esters according to preference, but it is probably better to disregard attempts to classify in this way since they are normally motivated by commercial more than chemical considerations.

It is clear from a consideration of the formation of the dye-fibre bonds that an inert compound is not formed. This can be shown best by considering two examples, (a) dyes based on heterocyclic systems and (b) dyes based on the vinyl sulphonyl system. If the following series of dyes is considered nucleophilic attack at the point indicated by the arrow is more ready in the order (a) < (b) < (c). In each case electron deficiency is created by the electron

(a)   (b)   (c)

attraction of the chlorine atoms, but this is supplemented and eventually dominated by the permanent influence of the nitrogen atoms. If the chlorine atom at the point of potential nucleophilic attack is replaced by CH_3O-, then the position is somewhat deactivated due to electron donation by the methoxy group. In the case of (a) the product will consequently be very resistant to attack by nucleophiles such as hydroxyl ions. In the case of (b) the resistance will be somewhat less and least in the case of (c). Thus the methoxyl group would be subject to hydrolysis in the same manner, but to a lesser degree, as the chlorine atom in the original compounds. If instead of a methoxy group, a cellulosyl group is used then the discussion relates directly to dye–fibre bond stability under alkaline conditions. Dyes of the type (b) and (c) are used commercially and their stability is adequate for technical purposes. Dyes of type (b) are not sufficiently reactive for several kinds of application in industry and attention has been paid to methods of increasing the extent of permanent activation. This has been by substituting the ring with electrophilic groups in the 5-position. Thus the 5-nitro analogues of (b) are very highly reactive and give very unstable bonds. The 5-cyano compounds are as reactive as the dyes of type (c) but exhibit sufficient permanent activation to give poor washing fastness. The 5-chloro analogues do not show very much increase in reactivity although the small advantage has been utilized. The 5-carboxy compounds are useful and when this group is used it is convenient to attach the reactive system to the chromophore through it, i.e.

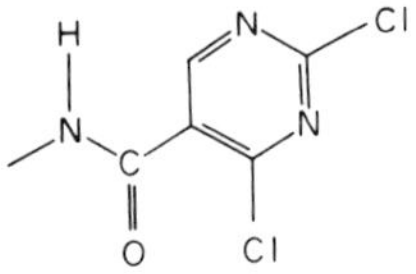

This device is useful because it provides a very attractive balance of properties and is employed commercially. (Dussy, 1970).

The majority of technically employed reactive dyes based on heterocyclic systems are covered by the following reactive systems.

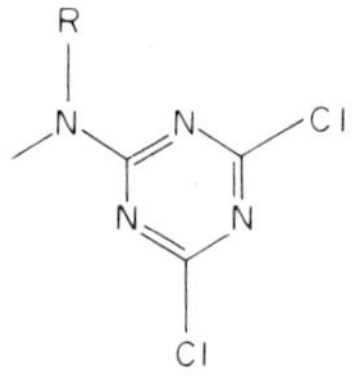

Procion M

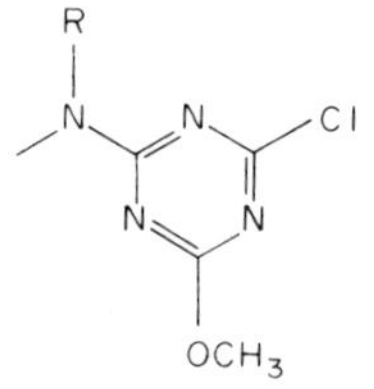

Cibacron pront

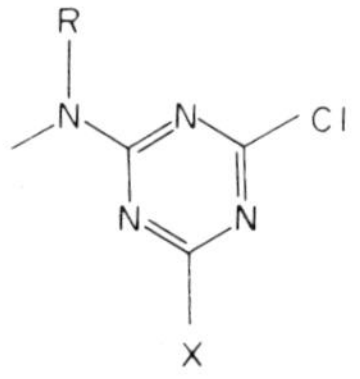

Procion H.
Cibacron

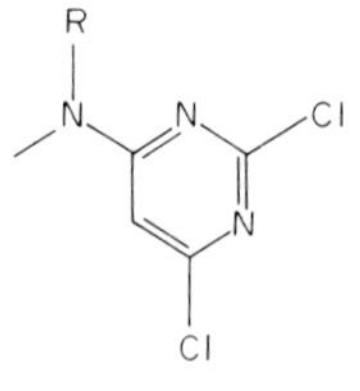

Drimarene,
Reactone

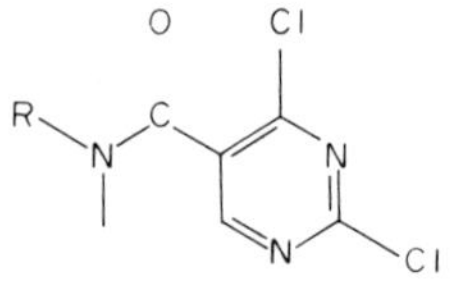

Drimarene R,
Reactofil

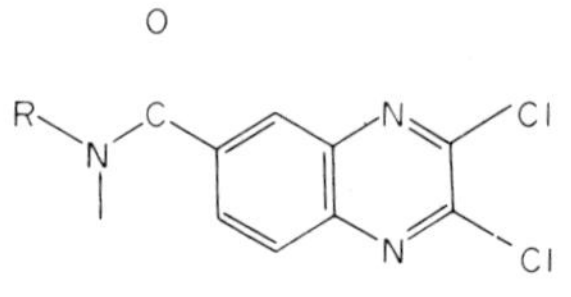

Levafix E

In the above, R signifies H or CH_3-, X may be a variety of "inert" groups, i.e. not replaceable under normal conditions of application or use. In all cases except the Procion H and Levafix E dyes, there are present two reactive centres giving rise to more than one possible fixation mode. Under most technical conditions this behaviour is significant only with the Procion M and Cibacron Pront dyes which can give the following fixation products.

Type 1

Type 2

Type 3

Clearly, Type 2 products may be obtained by further reaction of Type 1 products and Type 3 products may be obtained wither by direct hydrolysis of Type 1 or Type 2 products. Type 3 products exist exclusively in the keto form except at very high pH values. Type 1 and 2 exhibit the same permanent activation through the heterocyclic ring but differ because of the different inductive effects of the substituents.

Type 3 dyes show quite different permanent activation due to the isomerisation which breaks down the heterocyclic ring configuration. The result is that the three types show different bond stabilities under any given set of conditions.

In Type 1 dyeings both substituted positions (2- and 4- above) are susceptible to nucleophilic substitution. The electron deficiency at the 2-position is less than that at the 4-position so that the main alkaline hydrolysis reaction is Type 3 formation. But there is likely to be, under severe conditions, significant attack at the 2-position to produce significant dye–fibre bond rupture at the same time as hydrolysis. Under mild alkaline conditions Type 2 formation will predominate over hydrolysis as a reaction. Type 2 dyeings will be stable to alkaline hydrolysis since effective dye–fibre bond rupture requires two bonds to be broken simultaneously. When this does not occur and it is very improbable that it will, Type 3 formation will occur. In this case the permanent activation is much reduced and nucleophilic attack at the 2-position is much less favoured. The result is that Type 3 dyeings will be very stable to alkaline treatment. Since the hydrolysis conditions under discussion and the conditions of dye–fibre bond formation are comparable, dyeings produced from Procion M or Cibacron Pront dyes will be mixtures of Types

1, 2 and 3 with demonstrably different properties. The relative weaknessf the dye–fibre bond in the case of Type 1 dyeings is shown to some extent by the dye–fibre bond formed with Procion H or Cibracron dyes. In this case treatment with alkali cannot lead to deactivation by Type 3 formation so that the dye–fibre bond stability is less than it is in the case of Procion M or Cibacron Pront dyes, all other things being equal.

Acid-catalysed hydrolysis will proceed through protonation of the oxygen bridging group between the reactive system and the cellulose followed by nucleophilic attack by water. However, the different reactive systems show different behaviour because additional catalysis can occur because of further protonation in the heterocyclic ring itself or other reactions. In the case of the Levafix E dyes for example, direct hydrolysis of the imido bridging group between the chromogen and the reactive system occurs fairly readily. Something similar might be expected with the Drimarene R or Reactofil dyes but nothing has been reported. With the Procion H and Drimarene dyes acid catalysed hydrolysis, while clearly possible, is not technically important. However, due to the ease with which the heterocyclic –NH– group may be protonated to produce very high activation in the 2-position, acid hydrolysis of Type 3 dyeings can be important, and this has been studied in some detail.

Senn, Stamm and Zollinger (1963) studied the hydrolysis of dye–fibre bonds formed between cellulose and a number of fibre-reactive dyes based on structure VIII below.

VIII

R in VIII signifies a variety of reactive systems including those in Type 1–3 bonding. The pH conditions of hydrolysis were defined by a series of buffers. It was concluded on theoretical grounds that the hydrolysis under acid conditions was specifically catalysed by hydroxonium ions so that no unexpected anion effects could be expected although this had in fact been observed (Rattee, 1961). The comparative behaviour of the different reactive dyes was described in terms of % bond hydrolysis/hr by extrapolation of kinetic data. In general for all of the dyes examined a pH zone of maximum bond stability in the pH 6–7 region was observed and at any given pH value the relative stabilities were what might be expected from a consideration of the electronic distribution within the molecules

The study was continued by Pierce and Rattee (1967). Only type 3 dyeings were examined using simple azo chromogens. The results were again interpreted in terms of first order kinetics in which hydrolysis was catalysed by hydroxonium ions so that regular graphical analysis ($\log C_t$ against t) should give a rectilinear relationship from which the first order rate constant could be calculated. In fact the data gave results of the form shown Fig. 8.9, i.e. initial curvature passing into the expected linear relationship. This was interpreted

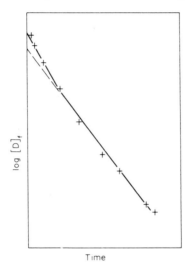

Fig. 8.9. Typical first order plot of dye–fibre bond hydrolysis data (Pierce and Rattee, 1967).

on the basis of the presence of two kinds of binding. The first and more reactive, it was proposed, represented terminal binding at the end of cellulose chains (Zollinger, 1960) while the second represented the expected binding on the 6-carbon atom. A number of other conclusions were drawn regarding the role of internal pH to account for the apparent effect of charge density.

It was later pointed out by Johnstone (1969) that contrary to Zollinger's hypothesis and the assumptions of Pierce and Rattee, first order kinetics should not be observed even if they truly describe the hydrolysis. The reason for this is that the catalysis occurs inside the fibre so that the internal kinetics will be governed by $[H_i^+]$. This will not, as implicitly assumed by previous workers, bear a constant relationship to the pH of the external solution, because the changing concentration of fixed dye during hydrolysis means a diminishing density of sulphonic acid groups in the fibre, and substantivity of buffer anions will lead to varying effects with different buffer systems. These criticisms were fully substantiated by experiment. The technique employed was more precise than that used before and enabled the kinetics of hydrolysis to be followed with precision. It was shown that the parabolic type of relationship observed by Pierce and Rattee was in fact curved throughout. The surprising fact was that all of the very considerable experimental data obeyed simple second order kinetics (i.e. $[C_t]^{-1} \alpha t$). Johnstone went on to show that all of the data of Pierce (1969) and of Senn, Stamm and Zollinger (1963) followed the same relationship. Further consideration suggests that the simple second order kinetic relationship arises from limited experimental discrimination.

Considering the situation inside the fibre and assuming specific proton catalysed hydrolysis, the rate of bond hydrolysis (dh/dt) will be

$$dh/dt = k[D]_t^F [H_i^+]_t \tag{8.35}$$

In order to explain the simple second order of relationship it is necessary to assume that

$$[H_i^+]_t = a[D]_t^F \tag{8.36}$$

where a is a constant, so that

$$dh/dt = ka([D]_t^g)^2 \tag{8.37}$$

and

$$ka_t = ([D]_0^F)^{-1} - ([D]_t^F)^{-1} \tag{8.38}$$

However, there is no reason whatever to suppose that (8.36) applies except as a crude approximation in specific cases. Calculation of the relationship between $[D]^F$ and $[H_i^+]$ using known data regarding the cellulose used applied to the Donnan theory showed that a slightly curved relationship existed. This was linear over limited ranges of $[D]^F$ values to a good order of approximation. Since in any experiment the value of $[D]^F$ changed by not more than 50% of $[D]^F$ it was possible to say that the relationship between $[D]^F$ and $[H_i^+]$ followed the empirical equation

$$[H_i^+]_t = m[D]_t^F + c \tag{8.39}$$

where m and c were functions of $[D]_0^F$.

Substituting (8.39) in (8.35) gives

$$dh/dt = km([D]_t^F)^2 + kc[D]^F \tag{8.40}$$

Integration gives

$$\ln\left[\frac{[D]_t^F}{m[D]_t^F + c} \right] \qquad \left[\frac{[D]_0^F}{m[D]_0^F + c} \right] \tag{8.41}$$

The values of m and c were obtained for any value of $[D]_0^F$ by computation so that the rate constant would be obtained graphically. The results showed that in the case of the two H-acid derivatives examined (C.I. Reactive Reds 1 and 2) the bimolecular hydrolysis constant was markedly dependent upon the internal pH and consequently upon those factors which affected that factor, i.e. depth of shade, ionic strength, external pH, buffer anion affinity, etc. The other dye examined (C.I. Reactive Orange 1) was relatively unsensitive in its behaviour to the pH conditions. The results are illustrated by Fig. 8.10.

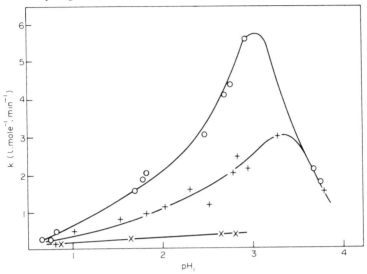

O C.I. Reactive red 1
+ C.I. Reactive red 2
X C.I. Reactive orange 1

Fig. 8.10. Variation of bimolecular fibre bond hydrolysis constants with internal pH (Johnstone and Rattee, 1973).

Dye–fibre bond unstability to acid conditions is found to be technically inferior with dyes derived from H-acid as compared with other Type 3 dyeings and the higher reactivity of the dyeings based on C.I. Reactive Reds 1 and 2 reflect this. So far no explanation for the effect has been put forward and data have not been produced over a more extensive range of dyes.

In the case of the vinyl sulphonyl reactive dyes on cellulose, the stability of the dye–fibre bond to alkaline conditions is the property which has received most attention. In the same way that sulphonic acid groups on the fixed dye lead to the condition $[H_i^+] > [H_s^+]$, there will be a lower concentration of hydroxyl ions inside the fibre so that the dye–fibre bonds will be more stable than expected. Nevertheless, the properties of the dye-fibre product will render it susceptible to attack by hydroxyl ions.

$$R-SO_2-CH_2-CH_2-OZ \xrightarrow{\quad OH^- \quad} R-SO_2-CH=CH_2$$

$$+ZO^- + H_2O$$

The important factor here is that the hydrolysis leads to unfixed *reactive* dye. If this does not exhaust back onto the cellulose then it will hydrolyse to the inert hydroxyethyl sulphonyl dye. However, under conditions where it will so exhaust it will fix onto the cellulose thus giving rise to permanent stains on white materials (Rattee, 1964). As a consequence of this dyes of this kind are required to exhibit low substantivity under likely washing conditions thus imposing a limitation on the choice of chromogens in research.

Finally, it should be remembered that the discussion of dye–fibre bond stability does not relate to the technical properties of commercial dyes except in a general sense. Commercial dyes are the end product of an extensive technical selection process so that in broad terms bad examples are eliminated. The physical and organic chemistry of any particular reactive dye system is mainly to be regarded as important to the research chemist in determining his success rate rather than to the ultimate user. That is not to say that the chemistry of the reactive dye systems does not provide a basis for the understanding of their use in practice.

II FIBRE-REACTIVE DYES ON NON-CELLULOSIC SUBSTRATES

Discussion so far has been in terms of dye–fibre fixation by nucleophilic substitution by cellulosyl ions. Other nucleophiles can of course be used, i.e. $-S^-$, $-NH_2$ etc., and this provides the basis for dye–fibre reaction with substrates other than cellulose, in particular nylon and wool. The general physical chemical analysis employed with cellulose would be expected to apply providing due allowance is made for real reaction possibilities since no mechanistic postulation is involved in the analysis. That is why the nature of the reactive system is not in principle important in such a consideration.

However, in considering reactions with wool a wide variety of reactions is possible and the situation is very complex. Shore (1968, 1969) has considered this problem in an important series of papers. A clear indication of the numerous reactions which it is necessary to consider is provided by the investigations of several researchers. The data of Table 8.6 are derived from that given by Shore (1968a).

It is clear that reaction with wool is complicated but no clear picture emerges from the data. Dyes bearing the same reactive system give a different balance of reactions. The data has been obtained in all cases by a method of amino acid analysis using dinitrofluorobenzene. This is known to give dubious quantitative values in some cases particularly lysine and cystine. Doubt has also been expressed about the data regarding reaction with histidine due to the instability of the dinitro phenyl residue (Craig and Konisberger, 1957). Other investigations have produced qualitative data regarding the sites of dye-fibre reaction. Silk has been shown to react with a

TABLE 8.6. Percentage substitution of amino acids in wool by different fibre-reactive dyes

Type of reaction / Substituted amino acid residue		Reaction with N-terminal residues							Reaction with mainchain residues						Reference
		Glycine	Alanine	Valine	Aspartic Acid	Glutamic Acid	Serine	Threonine	Cysteine	Arginine	Histidine	Lysine	Serine	Tyrosine	
Vinylsulphonyl dye (Remalan Brilliant Blue B)	Dyed at pH 6·2	65	42			40	30	33	24			27	27	16	Osterloh (1960)
Monochlorotriazinyl dye	"acidic"	23	27	27	23		27	27		13	12	1	24		Hornuff and Flath (1961)
Dichlorotriazinyl dye	"acidic"	45	46	47			44	44		23	23	2		30	Hornuff and Flath (1961)
Dichlorotriazinyl dye (Procion Brilliant Blue MR)	Dyed at pH 3·2-4·6								14		2	9	11	5	Hille (1962)
Monochlorotriazinyl dye (Cibacron Blue 3G)	Dyed at pH 3·6-4·0								13		-2	12	21	15	Hille (1962)
Monochlorotriazinyl dye (Cibron Blue 3G)	Dyed at pH 4·2-4·6								32		33	13	52	5	Hille (1962)
Vinylsulphonyl dye (Remalan Brilliant Blue B)	Dyed at pH 4·4-5·5								34		20	20	35	14	Hille (1962)
Vinylsulphonyl dye (Remazol Brilliant Blue R)	Dyed at pH 4·8-5·2								32		25	21	25	12	Hille (1962)
Vinylsulphonyl dye (Remazol Brilliant Blue R)	Dyed at pH 7·5-4·6								14		42	12	28	15	Hille (1962)

vinyl sulphonyl dye at the glycine, alamine, serine, histidine and lysine residues (Virnik and Chekalin, 1960). Derbyshire and Tristram (1960) have shown that acrylamido reactive dyes react almost exclusively with lysine residues. Lewis *et al.* (1965) have shown that ω-chloroacetylamino dyes react with thiol groups in cysteine residues, with histidine and lysine. Faced with this complexity, Shore (1968,b) examined the reaction of a dichlorotriazinyl dye with a series of model compounds which bore the same functional group as identifiable amino acids. The idea was to replace the peptide links by hydrogen atoms as shown by the following example,

—NH——CH——CO ——	H——CH——H	NH₂——CH——H	COOH——CH——H
CH₂OH	CH₂OH	CH₂OH	CH₂OH
Serine	Ethanol	Ethanolamine	Hydracrylicacid
	(i)	(ii)	(iii)

In the above (i), (ii) and (iii) are models for serine in the main chain, as an N-terminal and C-terminal residue respectively. The data for reaction with model compounds was taken to be representative of amino acid residues in proteins. This was checked with a number of soluble proteins of known constitution. A comparison between the reaction with the protein, a suitable mixture of model compounds and the calculated course of the reaction showed excellent correlation in a number of cases. A typical example is shown in Fig. 8.11.

The conclusions drawn from this study were that in sulphur containing proteins, thiol groups are important at pH values above 5. This confirmed earlier data (Lewis *et al.*, 1965). The lysine and hydroxylysine residues were shown to be the major basic reactive centres of interest.

Shore (1969) also describes the application of the general theory of reactive dyes discussed under cellulose to wool. This involved additional complexity since wool is not available in film form. Consequently Danckwert's equation for reactive diffusion into a semi infinite cylinder (equation 8.25) combined with Hill's equation for non-reactive diffusion were used in considering the kinetics of the reaction. The whole investigation provides a very fine example of meticulous logical analysis of a problem in terms of what is experimentally possible and succeeded in showing quite conclusively that the general theory of reactive dye fixation is applicable to wool–dye reactions. One important fact shown by Shore is the way in which the distribution of reacted dye residues varies with the degree of reaction.

The efficient reaction of fibre-reactive dyes with wool protein presents two main difficulties which are both reflected in less than optimum fastness to wet treatments. Firstly there is the problem of removing hydrolysed unfixed dye from the fibre at the end of the dyeing due to the generally high substan-

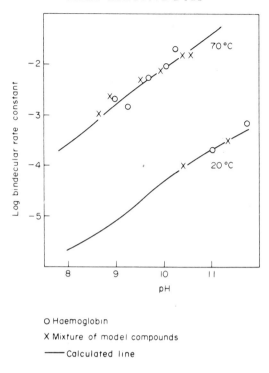

O Haemoglobin
X Mixture of model compounds
——— Calculated line

Fig. 8.11.   A comparison of calculated and observed reaction kinetics for a reactive dye/protein reaction (Shore, 1968b).

tivity of acid dyes for wool. Secondly there is the problem of dye–fibre bond stability. Reaction between a dye and an amino group in wool give a chemically stable compound and the dye should remain fixed to the fibre in so far as the latter is itself stable to hydrolysis. However, reaction between thiol groups and dyes does not give rise to stable compounds. The alkyl thiol derivatives of triazinyl dyes, for example, are very reactive. It would seem that this is not a readily solved problem and no reactive dyes display high wet fastness due solely to their covalent dye–fibre bonding. While reactive dyes are based on the formation of ester like compounds (in the case of wool, thioesters) this problem will remain. Possible ways around this difficulty have been demonstrated recently by Griffiths (1971) who uses quite a different approach to achieving dye–fibre reaction using thermal or photo-activation rather than chemical activation thus avoiding the reversible characteristics of conventional fibre-reactive dyes.

III REFERENCES

Aspland, J. R. and Johnson, A. (1965) *J. Soc. Dyers and Colourists*, **81**, 425, 477.
Beckmann, W. and Hildebrand, D. (1965) *J. Soc. Dyers and Colourists*, **81**, 11.
Beech, W. F. (1970) *Fibre-Reactive Dyes*, Logos, London.
Bhagwath, M. R. R., Daruwalla, E. H., Sharma, V. N., Venkataraman, K. (1970) *Text. Res. J.*, **40**, 392.
Craig, L. C. and Konisberger, W. H. (1957) *J. Org. Chem.*, **22**, 1345.
Danckwerts, P. V. (1950) *Trans. Far. Soc.*, **46**, 300.
Danckwerts, P. V. (1951) *Trans. Far. Soc.*, **47**, 1014.
Dawson, T. L., Fern, A. S., Preston, C. (1960) *J. Soc. Dyers and Colourists*, 76, 210.
Derbyshire, A. N. and Tristram, G. R. (1960) *J. Soc. Dyers and Colourists*, 81, 584.
Dussy, P. (1970) *Textile Chemist and Colourist*, **2**, 51.
Fern, A. S. and Preston, C. (1961) *Chimia*, **15**, 177.
Gardner, T. S. and Purves, C. B. (1942) *J. Amer. Chem. Soc.*, **64**, 1539.
Griffiths, J. G. (1971) *Belg. Pat. Spec.*, 765112.
Harned, H. S. and Owen, B. B. (1958) *The Physical Chemistry of Electrolyte Solutions*. 3rd ed., Reinhold, New York.
Hille, E. (1962) *Textil praxis*, **17**, 171.
Hornuff, G. von and Flath, H. J. (1961) *Faserforch und Textiltech*, **12**, 559.
Horrabin, S. (1963) *J. Chem. Soc.*, 4130.
Hughes, J. A. (1970) Unpublished work, University of Leeds.
Ingamells, W., Sumner, H. H. and Williams, G. (1962) *J. Soc. Dyers and Colourists*, **78**, 274.
Iyer, S. R. and Jayaram, R. (1970) *J. Soc. Dyers and Colourists*, **86**, 398.
Johnstone, J. E. (1969) M. Phil. Thesis, University of Leeds.
Johnstone, J. E. and Rattee, I. D. (1973) *J. Soc. Dyers and Colourists*, **89**, 89.
Lewis, D. M., Rattee, I. D. and Stevens, C. B. (1965) *Proc. Internat. Wool. Text. Res. Conf.*, Paris, CIRTEL 213.
Murthy, K. S. (1967) Ph.D. Thesis, University of Leeds.
Murthy, K. S. and Rattee, I. D. (1969) *J. Soc. Dyers and Colourists*, **85**, 368.
Neale, S. M. (1929) *J. Text. Inst.*, **20**, T373.
Osterloh, F. (1960) *Melhand Textilber*, **41**, 1533.
Pierce, J. H. (1969) Ph.D. Thesis, University of Leeds.
Pierce, J. H. and Rattee, I. D. (1967) *J. Soc. Dyers and Colourists*, **83**, 361.
Rattee, I. D. (1961) *J. Soc. Dyers and Colourists*, **77**, 739.
Rattee, I. D. (1963) *Amer. Dyes. Rep.*, **52**, 320.
Rattee, I. D. (1964) *Chimia*, **18**, 293.
Rattee, I. D. (1965) *J. Soc. Dyers and Colourists*, **81**, 145.
Rattee, I. D. (1969) *J. Soc. Dyers and Colourists*, **85**, 23.
Rhys, P. and Zollinger, H. (1966) *Helv. Chim Acta*, **49**, 760.
Senn, R. C., Stamm, O. A. and Zollinger, H. (1963) *Melliand textilber*, **44**, 261.
Shore, J. (1968a) *J. Soc. Dyers and Colourists*, **84**, 408.
Shore, J. (1968b) *J. Soc. Dyers and Colourists*, **84**, 413.
Shore, J. (1968c) *J. Soc. Dyers and Colourists*, **84**, 545.
Shore, J. (1969) *J. Soc. Dyers and Colourists*, **85**, 11.
Souchey, P. and Schaal, R. (1950) *Bull. Soc. Chim.*, France, 819.
Stamm, O. A. (1964) *J. Soc. Dyers and Colourists*, **80**, 416.
Sumner, H. H. and Taylor, B. (1967) *J. Soc. Dyers and Colourists*, **83**, 445.

Sumner, H. H. and Weston, C. D. (1963) *Amer. Dyes. Rep.*, **52,** 442.
Sumner, H. H. (1965) *J. Soc. Dyers and Colourists*, **81,** 193.
Virnik, A. D. and Chekalin, M. A. (1960) *Tekhnol. tekstil. prom.*, **19,** 109.
Zollinger, H. (1960) *Helv. Chim. Acta*, **43,** 1513.

Chapter 9

Problems of dyeing kinetics—specific situations

I THE TECHNICAL IMPORTANCE OF DIFFUSION IN DYEING

The relationship between the rate of diffusion of dye into a polymer substrate and the time it will take to complete a dyeing is fairly obvious. However in technical situations the processes involving diffusion of dye play an important part not only in relation to productivity but also to the evenness of the distribution of the dye in the system, i.e. the levelness of the dyeing. Frequently but not in every case the second factor is the more important.

In an ideal dyeing system the rate of supply of dye molecules to the adsorbing surface is uniform over the whole surface and ideally the same as the rate at which the dye can be adsorbed. Such a condition is sometimes to be approached in very small systems in a research situation but is never approached in general practice. Certain kinds of application in which the fabric is rapidly impregnated with dye solution so that the dye bath is entrained in the fibrous material can be said to come as near to the ideal situation as is possible in full scale operation but even here, as will be discussed, there are often serious problems.

The main difficulty with mixing or circulation dyeing systems involving dyebath exhaustion is that the diffusional situation is not uniform due to uneven liquor circulation, varying rates of liquor flow in different parts of the system or non-uniformity of the material to be dyed. The effect of these factors is enhanced because of the competitive situation, i.e. rapidly dyeing fibres leave insufficient dye available for more slowly dyeing fibres.

The effect of liquor flow rate on dyeing rates as a consequence of the hydrodynamic boundary layer has been discussed already (Chapter 3). It follows from this that under general conditions where dyeing rates are sensitive to flow, the dyeing rate will vary in a system where the flow rate is non-uniform. This adds to the obvious effect that in regions of high circulation dye is transported to the adsorbing surfaces more rapidly than in regions of lower circulation. When a dyebath is being heated during the dyeing operation, temperature variations will also lead to dyeing rate variations due to the relationship between dye affinity, the diffusion coefficient and the liquor viscosity with temperature as well as any possible effects on the adsorbing substrate.

The study of this kind of situation is essentially a problem in chemical engineering in which the mass transfer or mixing of dye in a system which can be regarded as a reactor. From an analytical point of view all dyeing machines involving liquor circulation may be regarded as equivalent but with specific design features which may affect the finer points of the treatment. One of the simpler systems to consider in this way is the hank dyeing machine in which suspended hanks of yarn are held while dye liquor is circulated by a paddle through a crude distribution system provided by a perforated plate. The dyeing system was treated by Burley et al. (1969) as a system of parallel rods (yarns) oriented parallel to a liquor flow in voids surrounding assemblies of rods. The dispersion or mixing process occurs perpendicular to the direction of flow towards the centre of assembly. The concept is illustrated in Fig. 9.1.

Mathematically the system is identical with the model for diffusion into a cylinder so that if a concentration gradient of a diffusant into the assembly of rods is measured a dispersion coefficient can be calculated. The variation of this parameter with flow can also be determined experimentally. In the study of Burley et al. a system with the dimensions of a carpet yarn dyeing system was used. The behaviour of an actual dyeing system was calculated by treating the system as one with simultaneous dispersion and adsorption and creating a computer analogue. The dyestuff properties were expressed by two parameters, the time of half exhaustion (τ) and the time of half levelling (T). The latter was defined as the time taken to transfer dye from dyed to an undyed material to that the concentration ratio was 2:1. It was assumed that the amount of dye adsorbed at any point in the system was directly proportional to the integral amount of dye transported to that point. The several time dependent simultaneous equations describing the processes operating in such a system can be solved on an analogue computer so that the analogue becomes a conceptual hank machine. The predictions of the model were shown to be remarkably accurate by direct experiment. The computer

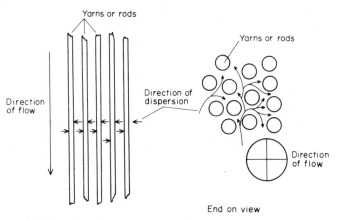

Fig. 9.1.   A schematic representation of hank dyeing.

output took the form of theoretical dyeing rate curves for different points on the hank radius as is exemplified in Fig. 9.2. Here the expected exhaustion–redistribution sequence in the outer layers of the assemblage of yarns can be seen clearly.

The term ϕ in Fig. 9.2 used to describe the time factor is dimensionless and is the ratio of real time to the time of half exhaustion. Thus the value of 6 for ϕ in Fig. 9.2 can mean a real time of 6, 20 or 30 minutes for dyes with times of half exhaustion 1, 3·67 or 5 minutes respectively. It is a useful device for extending the significance of data. It can be seen that there is a characteristic value of ϕ at which the depth of shade at the centre of the assembly approaches that on the outside. Burley *et al.* set the value of ϕ at which the ratio of depths at the outside and the inside was 1·05 as being the minimum time for hank penetration. In Fig. 9.3 this is shown as a function of the dispersion coefficient (and hence flow rate).

Clearly for any dyeing characteristics (defined by τ/T) there is a point at which the minimum hank penetration time is independent of the dispersion coefficient (and flow rate). Thus at sufficiently high rates of flow the uniformity of the dyeing will not be affected by variations in the rate of flow. By considering the known properties of dyes and technical requirements regarding acceptable dyeing times the range of dispersion coefficients and hence

flow rates which would be required in a dyeing machine was calculated. The
flow which would have to be achieved in a hank machine was shown to be
beyond that achieved in almost every commercial unit, certainly when
higher quality fast to washing dyes were involved because of the low τ/T
value of such dyes. Additionally as far as wool dyeing was concerned the
required flow rate was such that in a hank machine yarn damage was likely
to occur.

The particular somewhat negative conclusion to be drawn for hank machines is not necessarily valid for other kinds of machine. The general conclusion that it is possible and desirable to create a flow regime in which dye

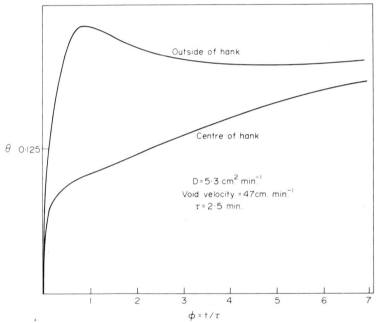

Fig. 9.2.   Rate of dyeing at the outside and centre of a hank calculated by analogue
computation (Burley *et al.* 1969).

mixing rates are independent of particular flow rates does seem likely to be
true for all machines, although the factors determining the desired flow rate
and the flow rate itself must depend upon the form of the machine and the
nature of the loading. It is a matter of common experience that dyeing rates
in all practical machines are dependent on the flow rate and it seems improbable that the optimum flow rate conditions (from the point of view of dye
mixing) are ever achieved. Thus in technical terms and until new kinds of
machine come available the mixing rate processes are determined by the
hydrodynamic factors in the machine. The improvement in the control of
dyeing kinetics must consequently depend on bringing the dyeing properties

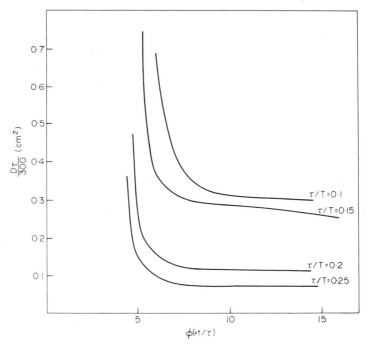

Fig. 9.3.   Dτ/300 as a function of dimensionless dyeing time over a range of dye properties and flow rate variations (Burley *et al.*, 1969).

(τ/T) of the system into line with circumstances as they present themselves.

For reasons quite separate from the rate of flow factors the control of dyeing behaviour is necessary in any case so that a consideration of the problem solely from the point of view of machine design developments cannot provide solutions to all the problems although it can provide a basis upon which the solution of the residual problems can be founded. Due to vagaries of manufacture, growth or pretreatment textile substrates often possess non-uniform dyeing characteristics. Changes in spinning, drawing etc cause changes in the dyeing behaviour of polyamide and polyester filaments, the natural life of sheep produce potential fibre-to-fibre dyeing differences in wool ("skitteriness") and on these as well as other materials non uniformity of pretreatments can produce variability which no amount of circulation control can cause to be dyed uniformily. The dyer is dependent entirely upon the dyeing properties of the dyes used under the particular dyeing conditions. In considering the flow rate factors the dyeing properties of a dye were expressed as the ratio τ/T which related the rate of redistribution of dye or levelling and the rate of dyebath exhaustion. It is by manipulating these two properties that the dyer may obtain a uniform result.

The redistribution properties of a dye arise from the reversibility of the adsorption. It is necessary consequence of the existence of adsorption and desorption in a steady state that even in a level dyeing dye molecules move from one part of the system to another by diffusional exchange within the substrate and exchange between adsorbed dye molecules and those in the external solution. In this context unlevelness may be seen as a localized distortion of the steady state situation. Given time natural redistribution will lead to a level result if the adsorption process is reversible. It follows from this description of the redistribution process that there must be unexhausted dye in the dyebath and that the rate of diffusion within the substrate will condition the rate at which dye may be desorbed. Without the former there can be no exchange with the solution and dye cannot be desorbed more rapidly than it can be transported to the fibre surface. These two factors also are the basis for the limitations of the redistribution phenomenon as far as producing level dyeings. It was shown by Lemin and Rattee (1949) that the rate of redistribution of dye was related to the amount of dye in the exhausted dyebath $(\tau \alpha (1-E)^{-1}$ where E is the exhaustion expressed as a fraction). The association of poor dyebath exhaustion with good level dyeing is well known in practice but the economics of technical operation limit the extent to which advantage may be taken of the relationship. The other problem with the redistribution process is that it is rapid only when diffusion coefficients are high with the consequence that wet fastness tends to be low. As dyes of higher fastness come to be used, their high exhaustion and slow diffusion largely preclude redistribution as a means to level dyeing and reliance has to be placed almost completely on controlling the rate of dyebath exhaustion.

Control of the rate of dyeing offers the advantage over reliance on redistribution as a means to levelness of being preventive rather than curative. In those cases where curative measures are simply not available prevention is clearly the only option. The normal means whereby dyebath exhaustion rates are limited, apart from the use of temperature control and circulation programmes in the dyeing operation, is the use of level dyeing assistants. These have been discussed in Chapter 4 and further consideration is given to the subject in Chapter 10. Further discussion is not necessary at this point.

Problems arising from diffusional factors in dyeing do not by any means arise solely from factors of non-uniformity of circulation, loading or substrate material. The extensive use of dyeing systems involving little or no circulation and a number of cases where dyeing rates require to be increased rather than reduced provide many examples of special problems which make useful case studies.

II SPECIFIC PROBLEMS IN COLOURATION DUE TO DIFFUSIONAL FACTORS

These general areas of interest will be considered relating to padding processes, the acceleration of dyeing rates and particular problems of dye compatibility.

A Non-circulation dyeing processes

These are processes in which no attempt is made to induce flow of dye liquor in the system. In some examples of non-circulating processes insufficient water is present for the concept of flow to be meaningful. The amount of dyebath solution present during the dyeing stage of the operation in processes of this type varies between 20% and 120% of the weight of material. The dyebath is thus entirely entrained in the material being dyed and it is normally applied by an impregnation process followed by uniform squeezing (i.e. padding) to ensure by mechanical means a general uniformity. In other cases the dyebath is even more localized, being applied by printing. Two aspects of such processes are considered here.

(1) *Tailing during padding processes*

In impregnation procedures there is inevitably a short but finite time between initial immersion and the squeezing when the material is in contact with an excess of dye liquor. During this very short time the material is wet out and time is available during which adsorption rate effects can become apparent. The effect during the passage of a small amount of material can be quite small but in continuous running the effect becomes cumulative so that any imbalance in the rate at which the different components of the impregnating liquor are taken up can become serious. During the impregnation stage the process is not free from circulation effects in a strict sense but this factor is not very important in a practical sense.

Differential adsorption of components in a padding or impregnation process when it leads to a change of shade along the length of the run is termed "tailing". The effect can be positive or negative, i.e. the shade may become stronger or weaker as the run proceeds, according to circumstances. During the 5–30 seconds immersion time during impregnation two general events may occur. With rapid diffusing dyes of high affinity sufficient exhaustion may occur to result in dye becoming removed more rapidly than expected. This effect will vary with the different component in a mixture so that the (negative) tailing which will occur may refer to a shade or a strength variation. When fibres to be dyed imbibe water more rapidly than the dissolved or suspended dye the opposite effect can occur leading to a strengthening of the impregnation liquor as the run proceeds. This results in a positive tailing effect. Positive tailing can be observed when material is impregnated with a pigment suspension. The pigment does not enter the polymeric substrate but water may, leading to the imbalancing effect. Certain substrates, e.g. viscose rayon, have a difficult to penetrate outer "skin". The slow diffusion of dyes through this can produce positive tailing effects.

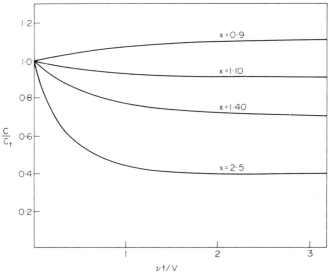

Fig. 9.4.   The depletion of pad liquors due to dye adsorption during immersion as a function of pad liquor changes.

The situation has been considered by Marshall (1955) who showed that broadly speaking real systems followed equation (9.1).

$$\frac{C}{C_f} = \frac{1}{x}[1 + (x - 1)\exp(-xtv/V)] \tag{9.1}$$

In equation (9.1), v is the rate at which padding liquor is removed from the trough by the material, V is the volume of the trough, t is the time from the commencement of padding and x is an affinity factor. The latter is defined as the fractional ratio of the dye concentration of dye on the material to that initially in solution. If no tailing effect occurs x is unity; it is greater or less than unity according to whether the tailing effect is normal or positive. By plotting the value of C/C_f against the number of changes of padding liquor in the trough (vt/V) the technical effect can be forecast as is shown in Fig. 9.4.

It can be seen from Fig. 9.4 that a steady state is achieved after about two pad liquor changes. This can be a considerable technical inconvenience. The kinetic effect can be limited by taking appropriate measures but the difficulty is that measures which reduce the rate of diffusion into the fibre increase the substantivity. A correction can be made by overfeeding dye to the pad trough, i.e. supplying dye at a higher rate to offset the exhaustion effect. The most effective method of avoiding the problem is by dye selection. Normally tailing effects are observed with cellulosic materials in which dyes exhibit a high rate of diffusion. Viscose rayon which behaves in many ways as if it has a difficult to permeate surface layer, wool which wets out slowly and has an epicuticle layer and synthetic fibres rarely show the effect in a serious way.

(2) *Diffusional control over dyeing rates in pad application processes*

In circulation dyeing processes the rate of adsorption of dye is generally dependent upon the rate of flow of dye solution past the adsorbing surface. Thus the rate of diffusion of dye through the hydrodynamic boundary layer is sufficiently slow for the dyeing rate to be affected by the thickness of the layer. When a dyeing process involves a completely unstirred dye liquor it is to be expected as a consequence of the known hydrodynamic boundary layer effects that the dyeing rate will be considerably slowed down. The situation in any dyeing process which involves no transport processes other than their diffusion must conform to the steady state equation (9.2),

$$\bar{D}_s \, \nabla_s = \bar{D}_f \, \nabla_f \qquad (9.2)$$

in which $\bar{D}_s$ and $\bar{D}_f$ are the diffusion coefficients of the dye in the solution and fibre phases respectively, ∇_s and ∇_f are the corresponding concentration gradients. It is normally supposed that $\bar{D}_f$ is so much less than $\bar{D}_s$ that the rate controlling process is diffusion in the fibre. However there is no means of knowing this by direct measurement. What is known is that if the diffusional transport process through the solution is supplemented by agitation or mechanical transport the dyeing rate may be greatly increased. Indeed in any particular case, the most rapid dyeing is achieved when mechanical transport is sufficiently great to virtually eliminate diffusional transport through the solution as a significant factor. In any research study of diffusion through films, fibres etc. it is precisely this situation that experimenters seek to achieve so that the rate determining process can be unambiguously defined as the diffusion in the substrate material. Thus all the circumstantial evidence in a large number of cases suggests that the diffusional transport process through the solution is likely to be rate determining in unstirred systems in a significant number of situations.

One important consequence of this analysis is the expectation that there will be a marked effect due to the actual volume of the entrained dyebath. Since this will be distributed over the surface of the material, the volume will be directly related to the "thickness" of the liquor and hence the distance over which diffusion has to occur. As a result the concentration gradient (∇_s) in the solution is affected producing an additional decelerating factor as the volume of the dyebath is increased. Normally in processes of this kind the volume of the entrained dyebath during the actual adsorption process is very low since the material is often dried after padding and dyeing takes place in a steamer. Thus the volume of the dyebath does not exceed the maximum inbibition value of the material to be dyed. This situation changed with the introduction of the pad-batch processes and the growing importance of this kind of procedure. The pad-batch procedure involves impregnating the material with between 70% and 100% of its weight of dye liquor and leaving it to stand in a batch for an appropriate period to complete exhaustion of the entrained dyebath. The batching stage may be at elevated temperature in a "pad-roll" machine which is effectively a steam heated storage chamber or more usually batching is carried out at ambient temperatures. The procedure has become particularly important in dyeing operations with the introduction of fibre reactive dyes and two examples of diffusional control in the solution phase are discussed in detail.

(a)   *The application of fibre-reactive dyes to cellulose by pad-batch procedures.*
Figure 9.5 shows a comparison between the rate of fixation of C.I. Reactive Red 1 applied by pad-batch and exhaustion/circulation methods. In both cases 10 g/l sodium carbonate was present as the alkali. The depth of shade was adjusted so as to give the same concentration of dye on the fibre in the two cases. It is clear that the pad-batch application gives much slower fixation and that circulation is important in relation to the fixation rate. With this very reactive dye the pad-batch fixation rate is largely liquor diffusion controlled. It is to be expected as a consequence that with highly reactive dyes fixation rates will vary as their diffusion coefficients in water rather than as their reactivity.

Dyes of lower reactivity would not be expected to exhibit this kind of behaviour and fixation rates would depend on reactivity. This is the case with monochlorotriazinyl reactive dyes, for example, which require 48–72 hours to complete fixation in cold pad-batch application. In such circumstances a clear advantage is to be expected from increases in reactivity up to the point where the rate of diffusion of the dye becomes the rate determining factor. Such increases have been achieved by the use of certain tertiary amines which act as catalysts through intermediate formation of the highly reactive quaternary amino triazinyl dye (Ulrich and Schaub, 1961) although the technical value of the effect is limited by compatibility problems (Dawson, 1964).

(b)   *The application of fibre-reactive dyes to wool by pad-batch procedures.*
The effect discussed in relation to dyeing cellulosic materials by pad-batch

procedures are to be seen particularly clearly when wool is dyed by the same method. Due to the non-uniform character of wool, the hydrophobic nature of the fibre surface and other properties it is necessary to adopt a padding solution which is fairly complex containing urea, a cationic wetting agent and a thickening dispersant, (Gibson, Lewis and Seltzer, 1970). Urea is present at a concentration of 300 g/l in the recommended process. At this concentra-

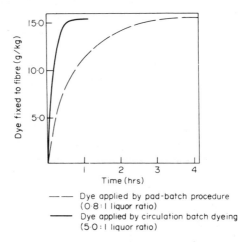

Fig. 9.5.   A comparison between the fixation rates of C.I. Reactive Red 1 applied with and without liquor circulation.

tion it has two effects, namely, to form a new compound by reaction with halogenotriazinyl dyes with higher reactivity (Swanepol, 1971; Gilchrist and Rattee, 1972) and to participate in a marked diffusional interaction with the dye (Tomita 1972) as shown in Fig. 9.6.

It is clear when allowance is made for the effect of diffusion rates in the entrained padding solution, that with highly reactive dyes the increased reactivity resulting from the presence of urea brings no advantage in fixation rates. As a consequence the discovery that an alternative method employing only 25 g/l urea brings no disadvantage in the form of lower dye reactivity and such a process is a practical proposition (Gilchrist and Rattee, 1973). With dyes of lower reactivity the opposite is found to be the case. When lower urea concentrations are used, the effect of urea on reactivity is not observed and fixation becomes very slow. With such dyes there is clearly "reactivity control" as distinct from "diffusion control".

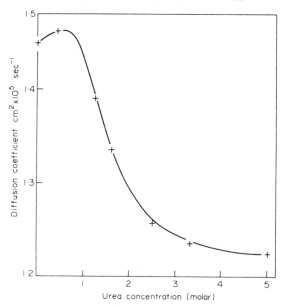

Fig. 9.6.   The effect of urea concentration on the diffusion coefficient of Acid Red 1 in aqueous solution.

B   The acceleration of the dyeing process

In the discussion of the diffusion process in Chapter 3 some consideration was given to the mechanism of the movement of a dye molecule through a polymer matrix. Since the latter is not a rigid arrangement the probability of its penetration by a fairly large molecule such as a dye will increase as mean distance between polymer chains increases. In many dyeing systems, particularly those involving natural polymers, the uptake of water provides sufficient competition for interchain bonding to allow the polymer to swell, i.e. to increase the mean interchain distance. The effect of water can also be enhanced by the use of higher temperatures to cause greater molecular motion. Thus cellulosic substrates which are not cross linked and take up water readily may be dyed easily from environmental temperatures upwards in the presence of water. Wool on the other hand swells less readily and offers greater diffusional resistance than, say, cotton. Adequate dyeing rates are achieved only at or near the boil as a consequence, when cross links are partly hydrolysed and interchain movement encouraged. However problems are presented when wool is to be dyed at lower temperatures, or with substrates which do not swell in water, e.g. polyester, triacetate, acrylic materials. With polyester fibres dyeing can be carried out under pressure at 120–130°C to produce adequate mobility in the polymer matrix but this is not without disadvantage and, in any case, cannot be safely employed with triacetate or acrylic materials. In these circumstances it is necessary to employ other means to bring about an acceleration of the dyeing process and two approaches have been extensively investigated.

(1)   The use of swelling agents

The role of water in producing swelling is one in which water molecules compete for interchain binding forces which contribute to the polymer matrix. Thus the movement of water into the polymer is accompanied by a increased polymer chain mobility and more ready diffusion by appropriate dye molecules. The same effect can be produced by other molecules providing they possess the necessary binding characteristics. To produce a significant effect quite considerable amounts of swelling agent are required to be adsorbed relative to the amounts of dye involved in a normal dyeing operation. The particular value of swelling agents is most obvious with hydrophobic fibres such as polyethylene terephthalate polymers which take up little water and normally swell very little in simple aqueous dyebaths. Compounds such as o-phenyl phenol or o-chlorophenol are strongly adsorbed in contrast and produce significant swelling action. Swelling agents do not need to be phenolic; diphenyl and chlorobenzene giving significant effects.

It is unlikely that as far as hydrophobic substrates are concerned there is any basic difference between swelling agents and disperse dyes except in so far as far greater quantities of the former are taken up by the polymer. The need for such quantities to produce significant swelling is not altogether surprising in view of the extent to which internal coehesive forces must be broken. The quantities of water required to produce polymer swelling in the case of hydrophilic fibres is correspondingly large. However water is fairly readily removed from dyed materials and has no toxic or sensitization effect on the skin as do many swelling agents. This imposes limits on the utility of many effective agents and makes removal of the swelling agent important.

The reversal of swelling is not rapid with some polymers such as polyethylene terephthalate with the consequence that the polymer may remain swollen even after swelling agent has been removed. The possibility of a reduction in fastness of the dyeing thus arises. Some swelling agents particularly β-naphthol have in addition a marked effect on the light fastness of dyes.

The search for dyes of very high fastness to washing and to sublimation has led to the increasing use of disperse dyes of very sparing solubility in water and the need to carry out dyeings at 120–130°C under pressure. Reference has been made already to the problems of crystal growth under these conditions (Chapter 7). In the presence of some swelling agents the solubility of some of these dyes is increased leading to more rapid crystal growth and instability of the dispersion. This has led to the recent introduction of swelling agents alleged to "stabilize the dispersion" at high temperatures. It is more probable that such agents are free from undesirable properties than that they possess positive stabilizing power.

Since they produce a change in the configuration of polymer molecules in the fibre, it is not surprising that swelling agents can have negative as well as positive effects on uptake rates in dyeing. It has been shown that many swelling agents effective in wool dyeing at low temperatures can reduce dyeing rates when present in small concentration (Fig. 9.7) (Cockett *et al.*, 1971) and this has been shown to be due to an allosteric competitive effect (Jayaram and Rattee, 1971). Thompson (1969) has shown that water has the same effect on the diffusion of azobenzene into cellulose acetate film (Fig. 9.8).

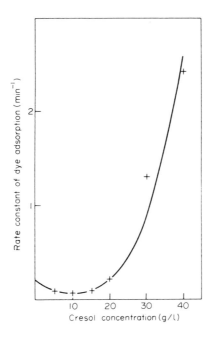

Fig. 9.7.   The rate of dyeing of C.I. Acid Red 1 at 25°C on wool in the presence of m-cresol (Cockett *et al.*, 1971).

(2)   *Solvent assisted dyeing*

The marked effect of certain organic solvents on the rate of uptake of dyes by wool in essentially aqueous dyebaths was reported by Peters and Stevens (1956) with the suggestion that lower temperature dyeing processes for wool dyeing might be employed. These observations paralleled studies by Karrhölm and Lindberg (1956) which were concerned with the effect of the wool cuticle on dyeing. Since that time solvent assisted dyeing has become a commercial process extensively used in polyamide dyeing as well as in the dyeing of wool. Technical aspects of the use of solvents in dyeing have been discussed by Swindell (1963) and by Beal and Corbishly (1971).

Solvent-assisted dyeing processes for wool dyeing involve carrying out the dyeing in the "aqueous dyebath containing a sparingly soluble polar solvent such as *n*-butanol, cyclohexanol, benzylalcohol etc. and a dye capable of

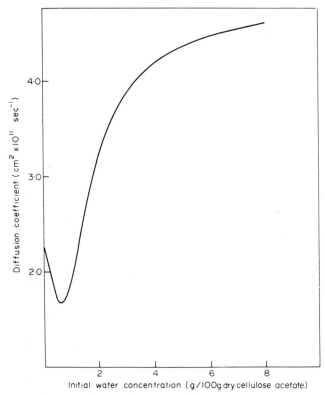

Fig. 9.8.   The diffusion coefficient of azobenzene in cellulose acetate in the presence of water (Thompson, 1969).

dissolving in both the water and the polar solvent. Other dyeing assistants, i.e. electrolytes, surfactants etc. may be present. Peters and Stevens found that wool could be dyed rapidly even at ambient temperatures by this method using appropriate dyes and solvents. Although the use of the process on a technical scale is well advanced the explanation of the effect is less clearly developed. There are reasons to suppose that the effect is in any particular case the result of several causes.

Reviewing several studies Beal, Bellhouse and Dickinson (1960) conclude that the following factors are operative in solvent-assisted dyeing,

solvents cause disaggregation of the dye thus increasing the activity of dye in solution

solvent is adsorbed at the fibre surface to give a solvent layer in which the dye is strongly dissolved thus leading to more rapid diffusion

the solvent brings about fibre swelling.

Peters and Stevens observed that the dyes most affected by the addition of solvent to the aqueous dyebath were the sparingly soluble milling dyes and that the method worked best when in a partition experiment the dye was more soluble in the solvent than in water. Dyes of this sort are generally strongly aggregated in aqueous solution. In many cases they are in a state of dispersion at temperatures below the boil. The addition of a solvent such as benzyl alcohol or *n*-butanol can be seen to bring about considerable disaggregation as evidenced by the intensity of colour and the clarity of the solution. There is no doubt that the disaggregating role of the solvent is capable of playing an important part in relation to the dye adsorption kinetics.

The theory that fibre swelling caused by the solvent participates in the acceleration of dye uptake is clearly inappropriate in the cases where useful solvents cause no swelling of the fibres, e.g. benzyl alcohol. Since benzyl alcohol and some other useful solvents can be effective then clearly swelling of the fibre is not absolutely necessary to stimulate rapid diffusion. The swelling effect of some other useful solvents, e.g. thymol may be a useful but not essential property in this context.

The hypothesis that a surface layer of solvent is involved requires careful examination. The useful solvents are all adsorbed by wool and polyamides in accordance with a BET isotherm. This indicates there is a distinct possibility of a capillary condensation or multilayer mechanism being involved. The concentration of bound solvent is generally a constant function of the relative saturation of the solvent solution showing that the energetics of adsorption and solution are very much the same. If a dyeing is carried out using solvent saturated with water as the dyebath solvent, very poor dyebath exhaustion is observed thus showing very low affinity for the fibre under such conditions. However under conditions of multilayer adsorption the volume of water-saturated solvents must be very small so that the liquor ratio relative to the saturated solvent dyebath must also be small. In such circumstances good exhaustion can be obtained even when affinity is low (cf. Chapter 10, Fig. 10.1). A combination of good exhaustion and low affinity is one which according to the pore model must be accompanied by a high diffusion coefficient and a rapid movement of dye in the fibre.

One of the predictions arising from this analysis is that the use of excess solvent will eventually lead to a dilution of the dye in the adsorbed multilayer so that the internal exhaustion will fall and the rate of dyeing becomes correspondingly lower. Certainly the behaviour of solvents is consistent with this situation.

It is probably more correct to regard dyeing as taking place from an adsorbed multilayer than to imagine the existence of a physically separable solvent layer. Indeed the best results of solvent assisted dyeing are to be seen before that situation can arrive. The distinction is important because the effectiveness of solvents can be demonstrated even in systems where at all times the solvent is in solution in the water. Swindell (1963) describes extensively the benefits of employing benzyl alcohol in dyeing carpet yarns at a concentration below saturation of the dyebath at all times. Saturation of the dyebath is, of course, unnecessary for the formation of adsorbed multilayers.

C   Some problems with specific substrates

Apart from the specific problems which have given rise to the topics already discussed, two substrates in particular have given rise to difficulties due to diffusional factors. Acrylic fibres have given rise to many problems due to their unusual dyeing characteristics particularly with regard to the difficulties of determining compatibility in mixtures of dyes. Human hair in cosmetic dyeing also presents problems of a special nature because dyeing must be carried out in the salon under conditions which do not cause more than a little discomfort to the hair's owner. These two topics are considered below.

(1)   Diffusion problems in the basic dye–acrylic fibre system

Acrylic fibres are generally dyed with basic dyes which on this substrate show high light fastness and brilliance of shade. However basic dyes exhibit high substantivity for acrylic fibres and very little redistribution or levelling behaviour. Uniform initial adsorption or strike of dye is consequently essential if even dyeings are to be obtained. The problems are compounded by the fact that in a mixture of dyes the dye uptake rate is a function of the relative amounts of the components of the mixture and the extent to which they are adsorbed. Thus non-uniformity of initial adsorption shows up as a variation in shade as well as strength and the dyeing of acrylic fibres with basic dyes involves the consideration of compatibility as well as uniformity of adsorption.

Harwood, McGregor and R. H. Peters (1972b) have shown that the diffusion coefficients of basic dyes in acrylic polymer films are markedly concentration dependent in the same way as occurs with anionic dyes on nylon films (McGregor et al., 1961). The diffusion coefficient increases with dye concentration in the film due to a lessening of the electrostatic frictional factor as the electrostatic saturation, θ, increases (cf. Chapter 3). The technical effect of this situation is such that when the initial adsorption of dye is uneven, the dye diffuses faster into heavily dyed fibres than weakly dyed fibres. Consequently a bad initial situation may well become worse before it gets better as a result of redistribution. With basic dyes on acrylic fibres the extent of dye redistribution is so little that it is essential to achieve even initial adsorption. Dyestuff chemists are consequently presented with the problem of producing ranges of dyes of uniform behaviour or finding suitable dyebath additives to control dyebath exhaustion. The problems are compounded by the need to establish a method of predicting compatibility.

In a particular case compatibility is easily assessed by a dyeing experiment. However in order to determine the compatible dyes in a range of twenty or more dyes the amount of work involved would be immense and each new speculative addition would involve an unacceptable amount of routine evaluation. The problem thus presents itself as one of predicting compatibility from characteristic dye adsorption parameters so that cross comparisons may be carried out easily. This problem arose first with the application of acid dyes to nylon and was considered by Atherton, Downey and Peters (1955, 1958). They showed that under conditions where the dyebath composition was constant (i.e. infinite dyebath conditions) if it was assumed that diffusion coefficients were independent of the concentration of dye in the fibre and the surface saturation of the fibre was unity then for a mixture of dyes, 1 and 2,

$$\theta = \theta_1 + \theta_2 = 1 \tag{9.1}$$

$$\frac{dQ_1/dt}{dQ_2/dt} = \frac{D_1 Z_1 C_{1,s}}{D_2 Z_2 C_{2,s}} \exp\left[\frac{-(\Delta\mu_1^0 - \Delta\mu_2^0)}{RT}\right] \tag{9.2}$$

where dQ/dt is the rate of accumulation of dye in the fibre, $D_{1,2}$ are the diffusion coefficients, $Z_{1,2}$ are the charges on the dye ions, $C_{1,s}$ and $C_{2,s}$ are the dyebath concentrations, $\Delta\mu_{1,2}^0$ are the standard affinities. Where dQ_1/dt and dQ_2/dt are equal it is clear that the dyes will be compatible in mixture.

The limiting condition of an infinite dyebath is not met in practice and the surface saturation does not equal unity at all times in most cases. However the problem can be solved for a finite dyebath to give a relation

$$\frac{\partial \ln C_{1,s}}{\partial \ln C_{2,s}} = \frac{D_1 Z_1}{D_2 Z_2} \exp\left[\frac{-(\Delta\mu_1^0 - \Delta\mu_2^0)}{RT}\right] \tag{9.3}$$

The conditions necessary for equation (9.3) to be true are that $D_{1,2}$ are constant and that θ_1/θ_2 is a constant at the fibre surface. Equation (9.3) can apply only when dyes 1 and 2 are compatible, $\ln C_{1,s} \alpha \ln C_{2,s}$ and θ_1/θ_2 is a constant. This means that as a further condition of compatibility

$$D_1 Z_1 \exp\left(-\Delta\mu_1^0/RT\right) = D_2 Z_2 \exp\left(-\Delta\mu_2^0/RT\right) \tag{9.4}$$

since equation (9.3) can be rewritten as

$$\ln C_{1,s} = \bar{K} \ln C_{2,s} + A \tag{9.5}$$

in which A is a constant and $\bar{K}$ is a *compatibility ratio* equal to the right hand side of equation (9.3).

Thus

$$\frac{C_{1,s}}{(C_{2,s})} \bar{K} = \exp A \tag{9.6}$$

Hence $\bar{K}$ must equal unity and equation (9.4) must apply.

Due to the insensitivity of log/log relationships it is not an appropriate test of compatibility to determine whether $\log C_{1,s}/\log C_{2,s}$ is a constant as Atherton, Downey and Peters demonstrate. Even incompatible mixtures are liable to show little curvature. However the use of the parameter $\bar{K}$ and equation (9.4) does provide both a description of the rate of dyeing behaviour of the dye and a means of describing compatibility.

Of course when D_1 and D_2 vary with θ then equation (9.4) cannot apply over the whole concentration range unless D_1/D_2 is a constant, i.e. both diffusion coefficients have the same relationship to θ and some consideration must be given to this condition.

The concepts of Atherton, Downey and Peters have been applied to the dyeing of acrylic fibres with basic dyes by Beckmann, Hoffmann and Otten (1972) and by Anderson, Bent and Ricketts (1972). This is possible because of the similarity of the soption and diffusion behaviour of basic dyes on acrylic fibres and acid dyes on nylon respectively (Harwood, McGregor and Peters, 1972a; 1972b). However in cases studied the condition that D_1/D_2 remains constant with θ does not apply since the slope of the relationship between $1/D$ and θ varies from one dye to another. Both sets of authors avoid the consequences of the predictable failure of the theoretical approach to describe compatibility by de-sensitizing the situation through the use of empirical tests to define the value of $D \exp(-\Delta\mu^0/RT)$ and using experience to determine technically permissable variations of $\bar{K}$. Beckmann *et al.* employ a dip test which uses a numerical factor related to $\bar{K}$ although it seems on the face of it that the use of a direct evaluation procedure such as dip test makes any theoretical analysis redundant. On the other hand Anderson *et al.* use a rate defining factor P/M^4 which is determined for single dyes without reference to actual dyeings of mixtures. The odd seeming parameter has theoretical justification.

The affinity of a basic dye DX_z may be defined as

$$-\Delta\mu^0 = RT \ln \frac{[C_f][X_f]^z}{[C_s][X_s]^z} \tag{9.7}$$

so that

$$\exp(-\Delta\mu^0/RT) = \frac{[C_f][X_f]^z}{[C_s][X_s]^z} \tag{9.8}$$

When two dyes 1 and 2 are present together in the same bath the anions are "shared" and since it is a condition of compatibility that $\theta_1 = \theta_2$ then the anion concentration terms may be ignored even when $Z_1 \neq Z_2$ and the compatibility equation (9.4) can be written as

$$D_1 \frac{[C_{f,1}]}{[C_{s,1}]} = D_2 \frac{[C_{f,2}]}{[C_{s,2}]} \tag{9.9}$$

Anderson *et al.* found that by using a suitable solvent the substantivity ratio $[C_f]/[C_s]$ could be replaced by a partition ratio (P) and the diffusion coefficient by a function of the cationic weight, (M^{-4}). Thus the parameter P/M^4 gives an indication of the magnitude of $D \exp(-\Delta\mu^0/RT)$. The precision cannot be high although there is some cancellation of errors but in technical terms the parameter enables satisfactory assessment of compatibility to be made and in any case dyers provide themselves with additional room for manoeuvre by employing anionic complexing agents in the dyebath to give additional control over dye uptake (cf. Chapter 4).

(2) The dyeing of human hair

Cosmetic colouration including the dyeing of human hair is a considerable industry. Although superficially the dyeing of human hair presents the same problems as the dyeing of other animal proteins special problems arise because of the particular nature of the substrate material and the unusual conditions under which dyeing is carried out.

Considering the latter first, hair is dyed on the head in a hairdressing salon. Consequently dyeing temperatures cannot exceed 40–50°C, dyeing times cannot be longer than 20–30 minutes, the chemicals must be physiologically safe, i.e. non-toxic, non-irritant, and then sensitizing properties must be at most very weak, and because of the general atmosphere of a hairdressing salon they must not give rise to complaints from other customers due to their smell. The dyeing operation is also carried out by hairdressers whose skill may be considerable but whose approach to chemical problems may well be rather different from that of a wool dyer.

Hair dyeing may be classified under three headings:

Temporary colouration is removable in a single wash and is not expected to last for more than a few days. The colour is often applied only for a special occasion and is often far from natural. Fastness is not a desirable property and diffusion of the colourant more than minimally into the hair surface is to be avoided.

Semi-permanent colouration is generally required to last for 6–8 weeks when the colour may be renewed or changed. Thus dyeing in the conventional sense is involved and consideration has to be given to diffusional problems. The shades involved are varied and may be fashion or natural shades.

Permanent colouration is generally favoured by the 35+ age group to cover greyness etc. The shades are consequently close to natural hair colours, i.e. browns, auburns and blacks. They do not wash out but have to be reapplied every three or four months to colour new hair growth.

Of these three only the last two need to be considered in any detail in the present context. In temporary colouration acid dyes (Milos and Brumner, 1965), basic dyes and metal complex dyes (J. R. Lyons, 1969) have been used and recently in a process avoiding fibre penetration altogether polymers containing chromophoric side chains have been employed (L'Oreal, 1968). The polymeric dyes are adsorbed in multilayers on the hair surface to give deep shades which are readily removed by shampoo detergents.

Semi-permanent and permanent colouration of hair requires some consideration of the diffusional problems. No hair dyeing process achieves a dyebath equilibrium. The requirement is to achieve a satisfactory degree of dye uptake in an acceptable time to produce a desired intensity of shade. The problem is that human hair is a highly cross linked structure. As a protein it has a high cysteine content which limits swelling and makes penetration difficult (Holmes, 1967). The relatively small openings which exist between cross links have been estimated to be about 0.6 nm in diameter (Bell and Breuer, 1971). Modification of the hair by swelling agents presents difficulties due to the sensitizing properties of most appropriate agents. The use of reducing agents to break disulphide cross links is undesirable because of the production of set and a modification of the hair which may not be required. Solvent-assisted dyeing has been explored on the basis of the work already described and enchancement of dye uptake rates has been observed (Cussler and Breuer, 1972). However the solvents tend also to degrease the human scalp so that the dyes stain the skin with a most undesirable effect.on the appearance. Thus the methods which have been adopted in the textile area with success to overcome problems of slow diffusion have not been open to or proved successful for the cosmetic dyer and it has been necessary to adopt special approaches based on new colouration methods or special selection of new dyes.

For semi-permanent colouration low molecular weight dyes of high extinction coefficient have been found. A balance has had to be found between the operation of low binding energies between small organic molecules and human hair (Breuer 1967, 1971) which tend to result in low fastness and the slow uptake of larger molecules. The nitro dyes amino and hydroxyl, derivatives of nitrobenzene (Tucker, 1971), and the amino and hydroxy derivatives of anthraquinone (Corbett, 1968) have been used extensively. The nitro dyes penetrate hair fairly readily but not so the anthraquinone derivatives. Their use is required however in order to achieve particular shades.

Permanent hair colouration generally involves the use of oxidation dyeing processes. These involve the initial adsorption of relatively low molecular weight molecules which may be combined in the hair to form pigment molecules by oxidative polymerization. The oxidation process is carried out by hydrogen peroxide or atmosphere oxygen. The full development of the colouration frequently involve the use of a colour coupler which reacts with the diamine formed by the initial oxidative process. The chemistry of the various elementary processes has been clarified considerably by Corbett (1969) and his general account of the reaction paths is shown in Fig. 9.9.

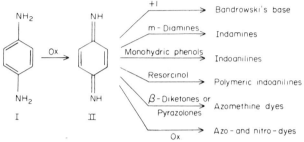

Fig. 9.9.   Reactive paths for Oxidation Hair Dyes (Corbett, 1969).

Typical combinations of oxidative bases and colour couplers is shown in Table 9.1 below which is also taken from Corbett's account.

TABLE 9.1.   Colours produced by oxidative dyeing on human hair (Corbett, 1969).

Oxidative base	Colour Coupler	Hair colour
p-phenylene diamine	None	Dark brown
	p-xylenol	Purple
	m-phenylene diamine	Blue
	2:4 diaminoanisole	Bluish purple
p-polylene diamine	None	Light red brown
	α-Naphthol	Violet blue
	Resorcinol	Blonde or brown

The great advantage of oxidation dyes is that during dyeing processes the hydrogen peroxide employed for oxidation simultaneously bleaches out dye previously deposited. This means that successive dyeings do not lead to a build up of pigment in the hair and there is a negligible difference between newly dyed and previously dyed hair. This is a point of obvious technical importance in relation to the dyeing of hair roots.

The process itself is somewhat unpleasant and sometimes leads to skin sensitization problems due to the use of diamines. Hydrogen peroxide also causes hair damage. Some improvements have been brought about by the use of pyridine derivatives to replace benzenoid compounds (Lange, 1965) but the products have not been adopted widely. The use of reactive dyes has been explored (Shansky, 1966) but it is clear from a consideration of the problem that special reactive dyes would be needed.

III REFERENCES

Anderson, W. L., Bent, C. J. and Ricketts, R. H. (1972) *J. Soc. Dyers and Colourists*, **88,** 250.

Atherton, E., Downey, D. A. and Peters, R. H. (1955) *Text. Res. J.*, **25,** 977.

Atherton, E., Downey, D. A. and Peters, R. H. (1958) *J. Soc. Dyers and Colourists*, **74,** 242.

Beal, W., Bellhouse, E. and Dickinson, K. (1960) *J. Soc. Dyers and Colourists*, **76,** 333.

Beal, W. and Corbishly, G. S. A. (1971) *J. Soc. Dyers and Colourists*, **87,** 329.

Beckmann, W., Hoffmann, F. and Otten, H. G. (1972) *J. Soc. Dyers and Colourists*, **88,** 354.

Bell, J. C. and Breuer, M. M. (1971) *J. Coll and Interfac. Sci.*, **37,** 717.

Breuer, M. M. (1967) *J. Phys. Chem.*, **68,** 2067.

Breuer, M. M. (1971) *Nature Phys. Sci.*, **229,** 185.

Burley, R. W., Rattee, I. D. and Flower, J. R. (1969) *J. Soc. Dyers and Colourists*, **85,** 187; 193.

Cockett, K. R. F., Kilpatrick, D. J., Rattee, I. D. and Stevens, C. B. (1971) *App. Polymer. Symp.*, **18,** 409.

Corbett, J. F. (1968) *J. Soc. Dyers and Colourists*, **84,** 556.

Corbett, J. F. (1969) *J. Soc. Cosmetic Chem.*, **20,** 253.

Cussler, E. L. and Breuer, M. M. (1972) *Nature Phys. Sci.*, **235,** 56.

Dawson, T. L. (1964) *J. Soc. Dyers and Colourists*, **80,** 134.

Gibson, J. D. M., Lewis, D. M. and Seltzer, I. (1970) *J. Soc. Dyers and Colourists*, **86,** 298.

Gilchrist, A. K. and Rattee, I. D. (1973) *Text. Chem. and Colourists*, **8,** 105.

Harwood, R. J., McGregor, R. and Peters, R. H. (1972) (a) **88,** 216; (b) **88,** 288.

Holmes, A. W. (1967) *J. Soc. Cosmetic Chem.*, **15,** 595.

Jayaram, R. and Rattee, I. D. (1971) *Trans. Far. Soc.*, **67,** 884.

Karrholm, M. and Lindberg, J. (1956) *Text Res. J.*, **26,** 528.

Lange, F. W. (1965) *Amer. Perfumer*, **80,** 33.

Lemin, D. R. and Rattee, I. D. (1949) *J. Soc. Dyers and Colourists*, **65,** 217.

L'Oreal, (1968) British Patent 993–181; French Patent 1484836.

Lyons, J. R. (1969) *U.S. Patent Spec.*, 3480377.

Marshall, W. J. (1955) *J. Soc. Dyers and Colourists*, **71,** 13.

McGregor, R., Petropoulos, J. H. and Peters, R. H. (1961) *J. Soc. Dyers and Colourists*, **77,** 704.

Milos, Z. B. and Brumner, W. H. (1965) *U.S. Patent Spec.*, 3164441.

Peters, L. and Stevens, C. B. (1956) *J. Soc. Dyers and Colourists*, **72,** 100.

Peters, R. H., Petropolous, J. H. and McGregor, R. (1961), *J. Soc. Dyers and Colourists*, **77,** 704.

Shansky, A. (1966) *Amer. Perfumer*, **81,** 23.

Swanepol, O. A. (1971) *App. Polymer Symp.*, **18,** 743.

Swindell, W. C. F. (1963) *J. Soc. Dyers and Colourists*, **79,** 457.

Thompson (1969) *Ph.D. Thesis*, Leeds University.

Tomita, M. (1972) University of Leeds, unpublished research.

Tucker, H. H. (1971) *J. Soc. Cosmetic Chem.*, **22,** 379.

Unilever Ltd., *Netherlands Patent Spec.*, 6700349.

Ulrich, P. and Schaub, H. P. (1961) *Textil Runst*, **16,** 815.

Chapter 10

Relating theory and practice

I THE PRACTICAL VALUE OF THEORY

It is often forgotten that scientific theories have a definite practical role in a particular context. With the development of experimental science came the acceptance of the idea that scientific theories should be verifiable in some appropriate practical way. Since experimental skill is not absolute and any measurement is subject to error there are limits to the extent to which a theory can be tested. In the foregoing chapters several examples have been quoted of "logically pure" theories which have been quite meaningless because many of the parameters involved were beyond the available experimental possibilities. This can mean that more than one scientific theory can exist at any one time when no experimental technique is available to distinguish between them. Both may subsequently be rejected in favour of a third theory but until the situation is resolved either theory may be used according to convenience. Sooner or later all scientific theories are shown to be at least partially false. Some are known to be false but continue to give good service because the accuracy of their predictions is quite satisfactory. Good examples of this are provided by the BET isotherm equation and the Gouy–Chapman equations for the distribution of ions near charged surfaces. Thus scientific theories are not concerned with absolute truth but are simply conceptual models which behave in the same way as the natural phenomena they purport to describe.

In applied science complications can arise because a wide range of standards of practice exist. The activities of so-called "pure scientists" tend to be comparable with regard to subjective standards but the context of the theoretical predictions in applied science may range from cosmetic dyeing in a hairdressing salon to a research laboratory in a department of physical chemistry. Thus a variety of conceptual models may be appropriate to deal with any broad situation. What must be determined is the approach appropriate to the circumstances. For example it makes little difference in a practical dyehouse whether disperse dyes are adsorbed by a solid solution or a site mechanism since both theories describe the practical situation equally well and enable the dyer to get on with his job. However the research chemist seeking new ways of inducing levelness in applying disperse dyes may well find one theory of far greater help than the other. On the other hand the practical dyer faced with crystal growth problems in his dyeing machinery is considerably helped in his appreciation of what can be done if he has some grasp of the dyeing mechanism and the properties of small particles.

A number of specific examples of the ways in which the application of physical chemical theory to dye adsorption phenomena have facilitated the solution of problems at different levels may be discussed.

II   EXAMPLES OF THE EFFECTIVENESS OF THEORETICAL UNDERSTANDING

Due to the broad spectrum of activities covered by any applied science, not least that which is the concern of the users of dyes, theoretical analysis may be used in many ways—

to rationalize empirical data and improve precision in development work

to consider the feasibility of proposed broad developments

to make cross disciplinary contributions in science

to assist day to day practice.

A number of examples relating to the physical chemistry of dye adsorption have been chosen which illustrate these points.

A Rate of dyeing and level dyeing assistants

When dyes are applied to substrates under industrial conditions lack of uniformity of packing or other causes of variation in the rate of supply of dye solution to the adsorbing surfaces will affect the process giving rise at some. stage in the operation to an uneven dyeing. In addition there may be differential effects in the substrate material itself so that the rate of dyeing of some fibres is greater than that of others. Thus in practice the production of a level dyeing depends upon a combination of controlled dyebath exhaustion to maximize the initial distribution of dye or "strike" and a procedure which permits reversibly bound dye to redistribute evenly over a period of time. It is known that the rate of redistribution of dye in an unevenly dyed load of material is related to the concentration of dye which remains in the dyebath (Lemin and Rattee, 1949). The higher this is the more rapidly will the distribution of dye become uniform. With dyes diffusing very rapidly in the adsorbant, the rate of redistribution can be so rapid that the uniformity of the initial "strike" is of little consequence. However dyes diffusing so rapidly must possess little fastness to wet treatments unless they are fixed in some way after the initial adsorption by some appropriate chemical means, e.g. dye-fibre reaction, metal complex formation etc. Thus with a large number of dyes used in practice a combination of even "strike" with moderate redistribution properties is employed. The faster to wet treatments the dyeing needs to be, the more important becomes the use of appropriate slow diffusing dyes and even initial adsorption.

It is possible to employ three kinds of dyebath assistant in the context of this general problem. There are *restraining agents* which have some effect on the rate of adsorption but primarily reduce the final exhaustion thus promoting redistribution. There are *retarding agents* which have no significant effect on the final exhaustion but slow down the rate of exhaustion. Thirdly there are *level dyeing agents* which minimize differential dye adsorption as between different fibres. The question which arises is whether the use of one kind of dyebath assistant is generally more useful than another and in a development programme which field is most likely to provide a high success rate. In order to answer this it is very useful to consider the physical chemistry of the process. Firstly it is clear that if different fibres take up dyes at different rates when availability factors, i.e. circulation etc., are equal, then reliance upon dye redistribution to produce levelness must prolong the dyeing operation quite apart from the disadvantages from the fastness point of view. Diffusion studies show, however, that when dye ions are involved reliance on redistribution may be misplaced for another reason. The work of Peters, Petropoulos and McGregor (1961) have shown that when anionic dyes are applied to polyamide fibres, the diffusion coefficient increases with the depth of shade or more correctly the concentration of dye in the fibre. Consequently the rate of flux of dye across the fibre surface increases as the dye concentration at the surface increases. Thus more heavily dyed fibres will take up dye more rapidly than weakly dyed fibres thereby enhancing problems caused by uneven initial strike. Peters *et al.* conclude from their work that agents promoting level strike will consequently be more generally effective than levelling agents in the dyeing of polyamide fibres and it may be presumed, wool or silk. This is a conclusion which would be reached otherwise only on the basis of expensively acquired experience.

One of the major developments in dyestuffs during the past twenty years has been the use of weakly polar 1:2 chromium complex dyes. These bear sulphonamido and related groups to confer dyebath solubility. Due to their high affinity and molecular weight, their rate of diffusion into protein and polyamide fibres is low, making the production of uniform dyeings almost completely dependent upon uniform strike and the absence of fibre-to-fibre dyeing differences. Thus with these dyes no element of choice existed as to which kind of level dyeing assistant might be the more effective. However two methods of attack on the problem remained. The dyestuff structures could be modified so as to give the required dyeing properties or special dyebath assistants could be developed. The role of the solubilizing group in dyes in controlling fibre-to-fibre differential dyeing effects had been clearly established by Townend and Simpson (1946) and this provided the basis for the design of suitable dye molecules ultimately based on weakly ionizing residues such as sulphonamido, methyl sulphonyl etc. A very clear account of this approach in which the physical chemical understanding of the dyeing process guided the organic chemist has been given by Schetty (1955). The alternative study of level dyeing agents also yielded valuable results which extended beyond the problem of the 1:2 chromium complex dyes for wool and essentially utilised the fact that dyes and surfactants interacted with one another in solution. This has been discussed already in Chapter 4.

The differential dyeing behaviour of dyes which dyestuff chemists endeavoured to avoid has been put to effective use by polymer chemists in the development of differential dyeing or deep dyeing man made fibres. It is not uncommon in dyestuff and dyeing technology for yesterday's problem to become tomorrow's asset.

B   Solvent dyeing

Quite apart from esoteric scientific interest in dye adsorption phenomena where the dye is dissolved in a non-aqueous solvent, the growing awareness of the need to exercise some control over the use of water and industrial effluents has stimulated both social and technical interest in the use of non-aqueous solvents for dyeing processes. It should be made clear that except for one or two regions of the world where water is often in very short supply, the main matter of concern is effluent control and a choice lies between the development of water treatment processes to enable the re-use of water and the limitation of undesirable effluents on the one hand and the use of a recoverable alternative solvent on the other. The present consideration excludes those special areas of interest where it may be possible to demonstrate absolute technical or economic advantages for the use of non-aqueous solvents in particular processes. While such areas exist, e.g. dry cleaning, it is fairly well accepted that in terms of dyeing operations in general non-technical advantages are needed to cause any inevitable change in the status quo. One important factor in these considerations is the feasibility and likely development cost of solvent-based processes and it is in this connection that a physico-chemical understanding of the dyeing process can play a vital part.

It is of course a prerequisite of any solvent-based process that the solvent should be cheap, freely available, safe to use, recoverable etc. In current terms this means the choice between trichloroethylene and perchloroethylene but the nature of the solvent is not relevant to the present discussion which is concerned with an analysis in basic terms of the dyeing process so that the role of the solvent may be isolated and considered.

In order for an organic chemical to constitute a useful dyestuff it must provide a satisfactorily strong coloured effect, i.e. its electronic energy absorption system at a molecular level must be fairly well developed. This means that in practice a useful dye will possess a molecular weight in excess of 350 and often in excess of 1000. Dye molecules are inevitably fairly large as a consequence. On the other hand useful textile materials in order to exhibit the necessary dimensional stability, strength, insulation properties etc., must possess appropriate properties at a molecular scale. When these two requirements are put together it is found that dyes are of such a size that they will diffuse through textile polymers only very slowly unless chemical or physical agencies are present to disrupt the cohesive forces in the polymer thus enabling diffusion to occur. The normal physical agency is heat while the normal chemical agency is a swelling agent, usually water. Thermal energy plays only a limited role except with thermo-plastic polymers such as polyethylene terephthalate or polyamide. In other cases it is necessary to have present an agent which can compete for the internal cohesive forces in the polymer thus promoting swelling which enables large molecules to penetrate. This role is played very efficiently by water in most cases, e.g. with natural fibres and some man-made fibres. In such cases the swelling caused by chlorinated hydrocarbon solvents is relatively slight so that a change of solvent introduces a special problem. With relatively hydrophobic fibres, e.g. polyethylene terephthalate and polyamide, the amount of swelling caused by immersion in water is small or negligible. Frequently special swelling agents.

such as chlorophenols are employed or very high dyeing temperatures are used to increase diffusion rates. With such fibres, chlorinated hydrocarbon solvents introduce no real disadvantage. From the point of view of promoting swelling they are little better than water so that no new problems are presented in that regard. Consequently there is a clear and theoretically predictable division in the problem.

At the present time, quite understandably, the greatest effort is expended in that area where success is to be most expected, with the hydrophobic fibres. This is assisted by the need to employ weakly polar dyes with some solvent solubility. It would be expected on theoretical grounds that the partition of the dye between chlorinated hydrocarbon and fibre would be favoured by using dyes which are somewhat more polar than can be used in aqueous systems and this is supported in practice. However dyeings still require to be fast to aqueous washing treatments and this inevitably imposes limitations. Quite considerable success has been achieved along these lines (Datye *et al.*, 1971) and numerous new dyes have been developed in anticipation of technical adoption of solvent dyeing.

Results obtained with natural and other hydrophobic fibres have been less successful for several reasons. The main reason is implicit in the system as has been described, i.e. the need for fibre swelling. This can be quite easily achieved by the use of phenols or water in the presence of chlorinated hydrocarbon solvents and good dyeings can be obtained. However their use introduces new problems and in any case the nature of the dye binding forces in the case of such substrates introduces a fresh problem not so far considered. The problems introduced by the use of water or phenolic swelling agents in the presence of chlorinated hydrocarbons mainly arise when it comes to solvent recovery. Firstly the use of water in excess of trace amounts demands the use of water-in-oil dispersing agents thus adding a costly contaminant to the system. Secondly water forms an azeotrope with the solvents and thirdly phenolic materials are often difficult to separate. Unfortunately it is not possible to avoid these difficulties because it is necessary to introduce a more polar solvent, e.g. water, into the dyeing system in order to achieve adequate transfer of dye from the dyebath. This is because the dye binding forces are hydrophobic interactions requiring the presence of water, or polar in character (dipole interactions) requiring a polar solvent to bring about dissolution of appropriate dyes and the operation of the electrical effects. The role of water in the operation of dye binding forces represents a second major basic advantage not possessed by chlorinated hydrocarbon solvents and this, combined with the need for fibre swelling, has so far stood in the way of generally attractive solvent dyeing processes for the dyeing of hydrophilic fibres. Ways around these difficulties may be found by the use of special kinds of fibre-reactive dyes or polymerizing dyes but experience suggests that it is generally very expensive and chastening to conduct research without a careful consideration of the underlying factors which are to be considered in physical chemical rather than technical terms.

C Dyeing theory and problems in polymer physics

It is because of their easy detection that dyes have such a useful part to play as indicators and coloured complexes in general analysis. The same properties render them potentially valuable in providing information about media on which they are adsorbed. In addition to their general coloured nature, dyes frequently exhibit dichroism so that the steric relationships between adsorbed dyes and polymers may be followed. Additionally the colour changes which sometimes accompany adsorption can be used to provide information about the substrate. In two areas in particular these properties are of growing importance, histology and polymer physics. Histology and biological section staining more often than not reflect ideas of dyeing which are rooted in superstition but in recent years physical chemical concepts have been applied to an increasing degree. Polymer physics on the other hand is a highly sophisticated subject deeply concerned with and contributing to the most advanced ideas of modern science.

Information regarding the state of orientation of molecular chains in polymers is a necessary part of their characterization and is essential to an understanding of the relationship between the physical properties and their structure. Normal methods of examination of orientation include optical birefringence, infrared dichroism etc. and these direct spectroscopic methods provide a great deal of valuable information. X-ray diffraction is also a standard technique. However these methods suffer from the shortcoming that technical polymers do not consist solely of crystalline material but considerable amounts of material in a varying state of order from crystalline to random. Direct spectroscopic methods provide effectively an integrated picture of the state of the polymer and consequently do not provide all the information that may be required. It is in this connection that dyes provide a useful means of obtaining a deeper insight into the situation. Dye molecules are too large to be occluded in or enter crystalline regions in polymers. On the other hand they may be detected when present in very small concentration so that their presence introduces no significant disturbance of the polymer chains. The study of the orientation of dye molecules in non-oriented or non-crystalline regions of polymers enables a study of those regions to be made separately from the crystalline regions. Detailed studies have been reported by Patterson (1954) and by Patterson and Ward (1957).

This approach has been carried further in diffusion studies by Blacker and Patterson (1969) when it was shown that the degree of orientation of adsorbed dye molecules in polyethylene terephthalate fibres increased with the degree of adsorption. From such data an orientation profile can be constructed for the polymer. Further possibilities in this connection have been shown by the work of Chantrey (1970). The tracer diffusion of a series of isotopically labelled azo dyes in polyamide film were examined. As the concentration of dye in the film was increased the diffusion coefficient increased markedly until a limiting value was reached. This was interpreted as meaning that the effective space in the polymer available for diffusion became fully occupied when the limiting dye concentration was reached. The possibility exists therefore of using dye molecules to determine free volume or rather effective free volume under actual dyeing or treatment conditions.

D   The application of fibre-reactive dyes to cellulose

The physical chemistry of the fixation of reactive dyes to cellulose is by no means completely understood. Nevertheless the degree of understanding is possibly greater than that for any other dye application. Remarkably accurate predictive equations have been developed to describe the physical chemical aspects of dye fixation (Chapter 8) and these may be used very effectively under practical production conditions in either a predictive or a trouble-shooting sense. From a practical point of view two properties are important in fibre-reactive dye application, namely the rate of dye fixation and the efficiency of fixation and the efficiency of fixation factor. These determine the productivity and the cost effectiveness of any application procedure. The equation describing these parameters have been discussed already in Chapter 8 and are repeated below.

$$\text{Rate of fixation} = [D]^f [\bar{D} . k_f . [OH]^s \exp\left(e\psi/kT\right) K'_{\text{cell}}]^{0.5} \quad (10.1)$$

$$\text{Efficiency factor} \quad \frac{[D]^f}{[D]^s}\left[\frac{\bar{D}}{K_h} . \frac{R'_{\text{cell}} . K'_{\text{cell}}}{[OH]^s} . \frac{\exp\left(e\psi/kT\right)}{[1+\exp\left(2e\psi/kT\right)]^2}\right]^{0.5} \quad (10.2)$$

Under practical conditions the operating variables, i.e. those over which control may be exercised directly are dyebath exhaustion, pH, temperature, reactivity and the ratio (W/W) of dyebath liquor to goods to be dyed (liquor ratio). In relation to these factors which may be designated E,pH, T,R and r respectively it is clear that

(I)   at any depth of shade $[D]^f$ is proportional to E,
(II)   the substantivity ratio, $[D]^f/[D]^s$ is equal to $rE/1-E$,
(III)   the factors involving the potential ψ both vary with the ionization of the cellulose. Thus $\exp\left(e\psi/kT\right)$ increases with pH while the more complex term in equation (10.2) decreases with pH,
(IV)   the terms k_f and k_h are both related directly to R,

(V) Since diffusion in cellulose follows the pore model (cf. Chapter 3) then $\bar{D}$ must fall as E increases.

Consequently the equations may be rewritten in terms of practical factors as

$$\text{Rate of fixation} \quad \alpha \quad E f_1 \left\{ \frac{\text{pH} \cdot T \cdot R \cdot}{E} \right\} \tag{10.3}$$

$$\text{Efficiency factor} \quad \alpha \quad \frac{r \cdot E}{1 - E} f_2 \left\{ \frac{1}{E \cdot \text{pH} \cdot T \cdot R} \right\} \tag{10.4}$$

The use of $f_1 \{ \quad \}$ and $f_2 \{ \quad \}$ signifies that the parameter depends upon some unspecified but direct function of the terms within the bracket. These two equations are general and conceal the interdependence of R and E with T and the way in which the function relating the diffusion coefficient with temperature is itself temperature dependent. Nevertheless within broad terms real systems behave as indicated by equations (10.3) and (10.4) and in terms of these it is possible to usefully modify processes to achieve the best results. The physical chemical analysis described in Chapter 8 can be applied through them to real procedures. Before showing how this may be done in some specific cases, it is necessary to show how the exhaustion, E, and the liquor ratio, r, are interrelated for a given substantivity ratio (S). This is shown in Fig. 10.1.

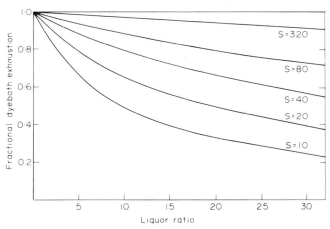

Fig. 10.1. The relationship between exhaustion and liquor ratio as a function of substantivity ratio.

It can be seen that when the liquor ratio is very low the exhaustion is very much less sensitive to the substantivity ratio than at higher values of r. On the other hand it is sensitive under such circumstances to small changes in r when the substantivity ratio is low. From the two equations (10.3) and (10.4) it is clear that high fixation efficiency may be expected when exhaustion is high, the liquor ratio is small, the pH and temperature are low and dyes of lower reactivity are used. On the other hand the rate of fixation would be expected to be low. In regular dyeing machinery the liquor ratio is normally 5–20 and the rate of fixation at ambient temperatures using low pH conditions and reactivity would be quite impractical. However by using padding procedures the volume of the dyebath can be reduced to so low a level that it can be retained by the material as entrained liquor. This reduces the liquor ratio to 0·5–0·8 and obviates the need to occupy a dyeing machine for long periods while fixation reaches completion. By operating the variables of pH and reactivity (this through dye selection) the dyer can then devise a wide range of processes, e.g. with highly reactive dyes,

$$\text{pad}\,(NaHCO_3) \quad \rightarrow \quad \text{batch (cold) 48 hours}$$
$$\text{pad}\,(Na_2CO_3) \quad \rightarrow \quad \text{batch (cold) 6–8 hours}$$
$$\text{pad}\,(NaOH) \quad \rightarrow \quad \text{batch (cold) 2 hours}$$

With less highly reactive dyes

$$\text{pad}\,(Na_2CO_3) \quad \rightarrow \quad \text{batch (cold) 48–72 hours.}$$

At the low liquor ratios employed, the exhaustion of the dyebath is relatively independent of the substantivity ratio so the effect of high pH on fixation efficiency is minimal. The same factor makes it pointless to add electrolyte to pad liquors to improve exhaustion but in cases where the substantivity ratio is rather small (i.e. low affinity dyes) it may reduce the sensitivity of the exhaustion to the liquor ratio and produce more reproducible results on a day to day basis.

The application of this approach to problems of batch dyeing in machines, i.e. where r has the value between 5 and 20, has been discussed already (Chapter 8). It is clear that the use of dyes of low reactivity presents many problems since high temperature, exhaustion and pH are necessary to produce a result in an acceptable time and it is not easy to arrive at a compromise situation with satisfactory efficiency unless dyes of high affinity are used. This can be done but introduces problems of wet fastness and unfixed dye removal. By careful dye selection manufacturers have been able to offer useful processes operating at higher temperatures using dyes with "middle range" reactivity. Efficiency in such cases is greatly enhanced by using two reactive systems or groups as in the case of the Procion HE dyes of I.C.I.

It is quite clear that if high speed fixation in a few seconds or minutes is required then reasonable efficiency will be obtained only when the liquor ratio is low. The first of the fibre-reactive dye processes for cellulose dyeing employed a procedure whereby dried material impregnated with dye was passed through caustic soda solution and steamed. The effective liquor ratio during the steaming operation was relatively high from the caustic soda impregnation liquor and the steam condensation involved in heating so that fixation efficiency was very low indeed unless the alkaline solution was saturated with salt. Even so the fixation was not high except when dyes of moderately high affinity were used, a factor which added further problems to washing off beyond those associated with the use of saturated salt solutions. Much better results were obtained by mixing sodium bicarbonate or sodium carbonate with the dye in the initial impregnation and then steaming the dried cloth. Although the fixation rate was lower at the lower pH, this was offset by the fact that the liquor ratio was very much smaller and heating rates were much higher. Thus fixation in a few seconds became possible. With highly reactive dyes fixation was significant during the drying of the impregnated cloth and the pad $(NaHCO_3)\rightarrow$ dry processes were developed. Both efficiency and rate of fixation depended greatly on drying conditions and associated wet cloth temperatures. In this the cloth structure, the type of heat input and the rate of heat input all had a part to play in accordance with expectations.

Developments in ranges of reactive dyes currently take the form of refinements designed to improve flexibility and choice. It would be true to say that the problems of dye stuff "design" in this context would be very much more considerable if the physical chemistry of the application of the dyes was not so well understood.

III REFERENCES

Blacker, J. G. and Patterson, D. (1969) *J. Soc. Dyers and Colourists*, **85,** 598.
Chantrey, G. (1970) Ph.D. Thesis, University of Leeds.
Datye, K. V., Pitkar, S. C. and Purao, U. M. (1971) *Textilveredlung*, **6,** 593.
Lemin, D. R. and Rattee, I. D. (1949) *J. Soc. Dyers and Colourists*, **65,** 217, 221.
Patterson, D. (1954) *Discussions of the Faraday Soc.*, **16,** 247.
Patterson, D. and Ward, I. M. (1957) *Trans. Far. Soc.*, **53,** 1516.
Peters, R. H., Petropoulos, J. H. and McGregor, R. (1961) *J. Soc. Dyers and Colourists*, **77,** 704.
Schetty, G. (1955) *J. Soc. Dyers and Colourists*, **71,** 705.
Townend, F. and Simpson, G. G. (1946) *J. Soc. Dyers and Colourists*, **52,** 47.

Author index

A

Subject index

A

Acrylic fibres,
 diffusion, 96
 dyeing, 292
Activation energy,
 of polymer crystallization, 99
 of dyeing, 44, 189
 of diffusion, 91
Activity of solutes, 118
Activity coefficients—determination, 118
Adsorption,
 activation energy, 28
 and diffusion, 87
 direct dyes, 182
 isotherms, isobars and isosteres, 30–38
 thermodynamic treatment, 42
 three-component systems, 40
 two-component systems, 27
 vat dyes, 211
Aggregation of dyes in solution, 122
 study by diffusion methods, 122
 study by spectrophotometry, 127
 study by temperature jump techniques,
  130
 effect of time, 127
Allosteric competition for binding sites,
 41, 161, 236
Amylodextrin—dye binding, 182

Amylopectin—dye binding, 182
Amylose—dye binding, 182

B

Basic dyes on acrylic fibres, 292
BET isotherm, 35, 41, 291
Binding forces, 2–25
Bovine serum albumen—dye binding, 158
Boundary layer, 109, 277, 284

C

Cellulose, 180, 245
 dye binding forces, 179–181
 dye saturation, 186, 192–193
 dyeing theories, 182–192
 dyeing with anionic dyes, 182–205
 dyeing with leuco ester dyes, 195–198
 dyeing with reactive dyes, 244–270
 dyeing with vat dyes, 198–219
 effect of chemical modification on dye
  uptake, 188
 internal volume, 187, 191–192
 ionisation, 245, 248
 reactivity, 246
 surface area, 186, 189

E

Einstein's diffusion equation, 98
Electrical double layer, *see* Diffuse
    electrical double layer
Electrokinetic effects, 15
Elovich equation, 30
Enthalpy of adsorption, 44
Entropy of dyeing with direct dyes, 182
Entropy of polymer crystallization, 99

F

Fibre-reactive dyes,
    alcoholysis and hydrolysis, 246–258
    effect of pH on exhaustion, 259
    hydrolysis of dye fibre bonds, 261–270
    isomerism, 259
    on non-cellulosic fibres, 271–273
    reactivity ratio, 247
    typical reactive systems, 264
    types of dye fibre bond, 265
Fick's Laws of diffusion, 51, 66
Free energy of adsorption, 2, 44
    of aggregation, 129
Free volume model of diffusion, 92, 96
Frenkels' equation, 28–29
Freundlich isotherm, 34
Frictional coefficient and diffusion, 98

G

Gibbs' equation, 52
Gibbs-Duhem relationship, 40
Gilbert cell, 147
Gilbert-Rideal treatment of wool dyeing,
    164
Glass transition in polymers, 92
    effect of diluents, 95

H

Hair dyeing, 295
    oxidative dyes, 297
Heat of dyeing,
    direct dyes, 182, 189
    of diffusion, 91
    of polymer crystallization, 99
Hydrodynamic theory of diffusion, 83
Hydrogen bonding, 19
    in dyeing, 181, 223
Hydrophobic dye binding,
    on proteins, 160
    on wool, 176
    with surfactants, 134
Hydrophobic hydration, 142
Hydrophobic interactions, 20

I

Ilkovic equation, 126
Internal pH, 11, 144
Isoionic point of proteins, 149
Isopiestic methods for determining
    activity coefficients, 120
Isotherms, 30
    effect of aggregation, 130
    additivity, 239

K

Kelvin's equation for vapour pressure at
    a curved surface, 203, 224
Kinetic theory of diffusion, 85

L

Laminar flow, 105
Langmuir isotherm, 32, 44